TO AMANDA AND ROBERT,
CHRISTOPHER, ALLEGRA, AND TIMOTHY

There are many reasons for studying the workings of the physical universe; we assume you have your reasons and a knowledge of elementary algebra including some trigonometry. Otherwise, very little is required as a prerequisite to this book.

TO THE READER

As the title suggests, our approach to an understanding of physics is through the simple behavior of ordinary objects. We do not seek to accumulate a multitude of results, but instead to trace out a few generalizations of considerable breadth and power. Rather than covering a lot, our strategy is to uncover the most important.

The text falls into five main areas:

The classical conservation laws—mass, momentum, and energy (Chapters 1 to 11)
The classical interactions—gravitational, electric, and magnetic (Chapters 12 to 16)
Light and waves (Chapters 17 to 20)
The physics of the very fast—relativity (Chapters 21 and 22)
The physics of the very small—quantum theory, atoms, nuclei, and elementary particles (Chapters 23 to 28)

In developing these topics we have tried to choose arguments that are concise, yet inescapably clear. These conditions are frequently met by the actual historical approach—for example, Christian Huygens' introduction of energy.

A word about some special notations used in this book. The symbol **+** marks a section that is *extra*, in the sense that it is an interesting extension of the main theme but is not essential to the continuing story. The symbol **c** marks other extra items for which *calculus* is required. A reader unfamiliar with calculus may simply skip this material.

Examples worked out and explained in detail are set off from the main body of the text by a shaded background.

For help in preparing "Physics from the Ground Up" the authors are indebted first of all to students and instructors at Rutgers College who used and contributed to the improvement of preliminary versions. We are also indebted to Mr. David A. Beckwith of McGraw-Hill, who contributed suggestions and criticisms in unusual measure. Our special appreciation goes to Professor Robert Ehrlich, who prepared the computer-produced electron-density pictures appearing in Chapter 26. One of us (HYC), whose lecture notes comprised the first version of this book, owes a special debt of gratitude to three teachers: Professors Sergio De Benedetti, Gerald Holton, and Edward M. Purcell.

HERMAN Y. CARR
RICHARD T. WEIDNER

CONTENTS

THE
ELECTRIC
INTERACTION

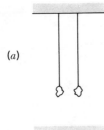

(a)

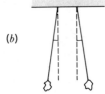

(b)

FIGURE 13-1 (*a*) Two small pieces of plastic sheet suspended side by side from light threads. (*b*) After being rubbed with wool they repel each other.

The electric force is enormously greater than the gravitational force.

Notably different from gravitation is the second basic force, the electric force. Perhaps as early as 600 B.C. the Greeks knew that *elektron* (amber), after being rubbed, attracts bits of straw or feathers. We can observe the same effect between a wool sweater and a sheet of plastic. The fact that no attraction appears until the two objects have been rubbed together at once proclaims it to be a new kind of force. It must arise from some other charge than gravitational mass, since we detect no change in the mass of either object after friction.

This new interaction displays a property we never found for gravity: the force can be repulsive as well as attractive. If two pieces of plastic are hung side by side on threads, as shown in Fig. 13-1, they will flee from each other after being separately rubbed with wool. Especially when the weather is dry, it is easy to make the threads deviate from the vertical by a sizable angle, 10 or 20°. From this fact it can be inferred that the horizontal repulsive force on each piece is comparable in magnitude to its weight. (For a deflection of 45° the two forces would be exactly equal in magnitude. Why?) But the weight of a piece is the gravitational pull of the entire *Earth* upon it. This means that the repulsive force between two pieces of plastic is enormously greater than the gravitational attraction between them. Taking the mass of the Earth as 10^{25} kgm and the mass of the piece of plastic as less than 1 gm, we conclude that the electric force in this situation is more than 10^{10} times stronger than the gravitational force.

It is hardly necessary to stress the importance in our lives of this potent and ambivalent force. The very workings of our body—from the pressure exerted by our hand against a door to the functioning of the billions of interconnected nerve cells of our brain—are manifestations of electric force.

13-1 THE CONCEPT OF ELECTRIC CHARGE

Electric charge, the intrinsic property of an object which governs its electric interaction with other objects, must now be defined in an operational way. The official procedure, for the mks system, is rather complicated, involving moving test bodies and the magnetic effects between them. Since magnetic effects will not be treated until Chapter 14, we shall provisionally define electric charge by the same sort of experiment used for gravitational charge. That this is possible is our first indication of the similarity between the two. Heretofore qualitative differences have been emphasized; but as far as their mathematical description is concerned, the electric and gravitational interactions are much alike.

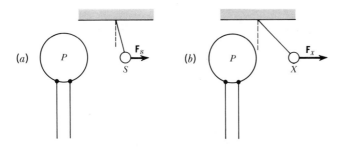

FIGURE 13-2 Determination of the electric charge of an object. (*a*) A standard object *S* of electric charge q_S is brought to a location near a fixed object having a large electric charge, and the interaction force $\mathbf{F}_S$ is measured. (*b*) Then the object *X* of unknown electric charge q_X is brought to the same location, and interaction force $\mathbf{F}_X$ is measured. By definition the value of q_X is given by [13-1].

Recall the procedure for gravity (Section 12-1). Two objects were weighed at the same point of space, and the weight ratio was taken as the charge ratio. In the electrical case, we can replace the massive Earth by a highly charged body, say the large metal sphere of a Van de Graaff generator (see Addendum 2 of Chapter 14). First a standard electrified object *S* then an arbitrary object *X* is brought to the same location relative to the sphere, and the forces $\mathbf{F}_S$ and $\mathbf{F}_X$ of the two interactions are measured (see Fig. 13-2). The ratio of the unknown electric charge q_X to the standard electric charge q_S is defined to be

Electric charge magnitude defined

$$\frac{q_X}{q_S} = \frac{F_X}{F_S}$$

[13-1]

Let's get an idea of the magnitudes involved in [13-1]. The mks unit of charge, the *coulomb,* happens to be (as magnetically determined) enormous. If our standard body *S* were a metal sphere 1 cm in radius which had been touched to an ordinary 90-V battery (Fig. 13-3), it would carry a charge of $q_S \approx 10^{-10}$ C, a typical charge by laboratory standards. The force between it and a Van de Graaff sphere 1 m away would be sizable: $F_S \approx 1$ N. We can now appreciate how big a coulomb really is. If q_X were anywhere near 1 C, the measured force would be about 1 million tons!

Electric charge unit, the coulomb (C)

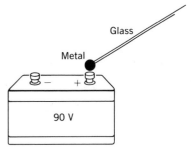

How do electric charges add, i.e., what is the total charge of a composite system consisting of several independently charged objects? The answer, although simple, is not obvious; it comes from experiment. Suppose that we take one object having a charge of magnitude q_1 and another object with a charge magnitude q_2 and tie them together (being careful, of course, to handle them only by means of insulating strings, so that the charge is not lost) and then measure the charge on the two tied together (by measuring the electric force on the composite object; see Fig. 13-4). The result is this: the charge of the composite object is *sometimes* just the arithmetic sum of q_1 and q_2, but not always. The composite object has a charge

FIGURE 13-3 One procedure (tentative) for obtaining a standard electric charge: a 1-cm metal sphere is touched to the terminal of a 90-V battery (the usual procedure will be discussed in Chapter 15).

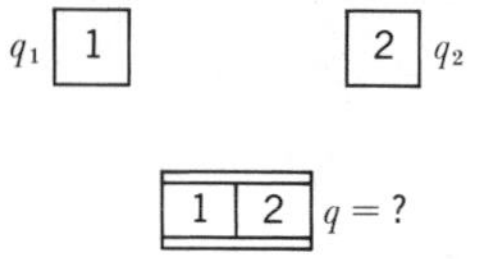

q_1 | 1 | | 2 | q_2

| 1 | 2 | $q = ?$

FIGURE 13-4   A composite charge q formed by fastening the individual charges q_1 and q_2 together.

Electric charges add algebraically.

Two kinds of charges: like repel; unlike attract

[1] The electrical neutrality of all ordinary objects actually shows that the electric force is strong. Since there are very closely equal amounts of positive and negative charge, and since the two charge types strongly attract each other, no object can easily avoid being electrically neutral.

Electron charge: -1.6×10^{-19} C

of magnitude $q_1 + q_2$ only if the forces of the fixed object on q_1 and q_2 individually are either both repulsive or both attractive. On the other hand, if the force on one object is repulsive while the force on the other is attractive, then the net, or total, charge for the composite object is found to be the arithmetic *difference* $q_1 - q_2$.

We can account for both situations by associating an algebraic sign with electric charge. The two algebraic signs, positive and negative, can be used to designate the two kinds of charge, the one kind attracted by our fixed object and the other kind repelled by it. More important, if the charge is considered to be an algebraic quantity having both magnitude and sign (kind of charge), then the charge of a composite object is always the *algebraic* sum of the individual charges. The sign of the charge automatically gives the arithmetic sum or difference, whichever is appropriate. For example, taking the arithmetic difference of the magnitudes 5 C and 2 C for two objects of unlike kind is the same as taking the algebraic sum of $+5$ C and -2 C. Using either procedure, we find the total charge to be $+3$ C.

We have identified two kinds of charge: one kind is attracted by our fixed sphere, the other repelled. It is now an easy matter to establish experimentally that objects having the *same sign* (kind), positive or negative, *repel* each other, while objects having charge of *opposite sign attract* each other.

Choosing the sign convention, i.e., associating one of the two kinds of charge with a positive or negative sign, is an arbitrary matter. The convention now universally used was first introduced by Benjamin Franklin. By Franklin's convention, an uncharged object touched to the terminal labeled $+$ of an ordinary battery becomes positively charged, and conversely for the negative terminal.

Most ordinary objects are electrically neutral, i.e., have zero total charge, at least insofar as we can tell with simple measurements.[1] This implies that the atoms of neutral objects are neutral too. The charge of any one particle in the atom is very small. For example, an electron has a charge $-e = -1.6 \times 10^{-19}$ C and the proton $+e = 1.6 \times 10^{-19}$ C (the neutron, of course, is neutral). Thus, an object with only a modest charge of 10^{-10} C has an enormous excess of particles having charge of one sign. Then how were we able to assert that giving an object an electric charge did not appreciably change its mass?

The answer lies in the smallness of the *fraction* involved. Although a huge number of particles may be transferred to or from an object when it is touched to a battery terminal, this number (even if it is as large as 10^9, roughly the number of electrons in 10^{-10} C) is minute compared to the total number of atoms in a small metal sphere, which is of the order of 10^{23}.

Moreover, it has been established that in normal metals charge is transferred by the movement of electrons, and an electron is less massive than a proton or neutron by a factor of nearly 2,000. The change in mass arising when an ordinary object is touched to the battery terminal is no more than 1 part in 10^{17}, an undetectable difference.

Query Can you confirm that 10^{23} is approximately the correct number? (Recall what you know about kinetic theory and the relative densities of gases and solids.)

A small charged object, having a 3-gm mass, is placed 1 m from the center of the large charged sphere of a Van de Graaff generator, as shown in Fig. 13-5. The thread by which the small object is suspended makes an angle of 10° with the vertical. What is the electric force exerted on this object by the charged sphere?

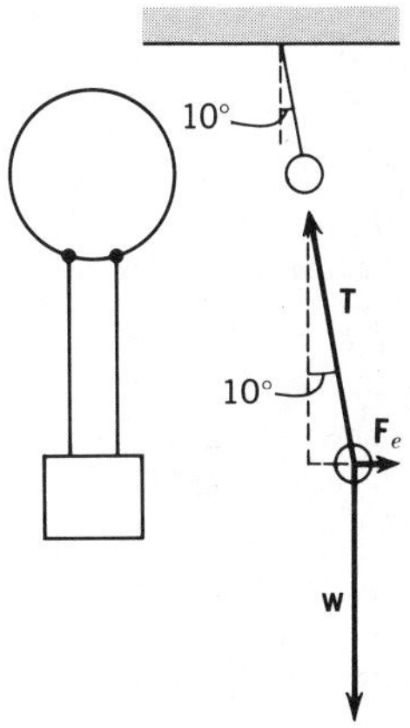

EXAMPLE 13-1

The object's weight is $w = mg = (0.003 \text{ kgm})(10 \text{ m/s}^2) = 0.03$ N. Since the object is in equilibrium, the vertical component of the force of the thread is equal in magnitude to this weight

SOLUTION

$$T \cos 10° = w$$

Moreover, the horizontal component of this force is equal to the electric force

$$T \sin 10° = F_e$$

Eliminating T, we have

$$F_e = \frac{w}{\cos 10°} \sin 10° = w \tan 10°$$

$$= (0.03 \text{ N})(0.18) = 5.4 \times 10^{-3} \text{ N}$$

FIGURE 13-5

What is the net charge on the small object in Example 13-1 if a standard charge of 10^{-10} C experiences a force of 1.8×10^{-6} N when placed at the position of the small object (see Fig. 13-5)?

EXAMPLE 13-2

Since we are given the force F_S on the standard charge of $q_S = 10^{-10}$ C, we compute the unknown charge q_X from the ratio [13-1] of F_X and F_S

SOLUTION

$$\frac{q_X}{q_S} = \frac{F_X}{F_S}$$

or

$$q_X = \frac{F_X}{F_S} q_S = \frac{5.4 \times 10^{-3}}{1.8 \times 10^{-6}} 10^{-10} = 3.0 \times 10^{-7} \text{ C}$$

EXAMPLE 13-3   A piece of laboratory apparatus of ordinary size may typically acquire a charge of 1 microcoulomb (μC, 10^{-6} C). (a) If the charge is negative, how many additional electrons has the object acquired? (b) What is the increase in its mass as a consequence? (The electron mass is 9.1×10^{-31} kgm.)

SOLUTION   a   The number of electrons is

$$\frac{10^{-6} \text{ C}}{1.6 \times 10^{-19} \text{ C/electron}} = 6.3 \times 10^{14} \text{ electrons}$$

Although enormous, this number is small compared to the total number of electrons in a cubic centimeter of solid material, about 10^{23}.

b   The increase in mass is

$$(9.1 \times 10^{-31} \text{ kgm/electron})(6.3 \times 10^{14} \text{ electrons})$$
$$= 5.7 \times 10^{-16} \text{ kgm}$$

a truly negligible mass compared to that of laboratory apparatus.

13-2   ELECTRIC-CHARGE
CONSERVATION

Again suppose that a neutral piece of wool and a neutral piece of plastic have been rubbed together. These two charged bodies, suspended on insulating strings, are allowed to make contact and as a single joined body are subjected to the measuring procedure of the previous section. The measured charge is found to be zero. *The electric charge of an isolated system is conserved.* This generalization does not, of course, rest on a single test. In Section 14-4, where the properties of conductors are discussed, we describe extremely sensitive procedures capable of detecting violations of charge conservation; none has ever been found.

Electric-charge-conservation law: total charge of an isolated system is constant.

To underline that the conservation principle applies only to *total* charge, we cite two phenomena from the domain of subatomic physics. In so-called *pair annihilation* a negatively charged electron and a positively charged positron (same mass as the electron and same charge magnitude) destroy each other and produce two electrically neutral highly energetic gamma rays, or particles of electromagnetic radiation (see Fig. 13-6). The individual charges have vanished out of existence, but the total charge is preserved as zero.

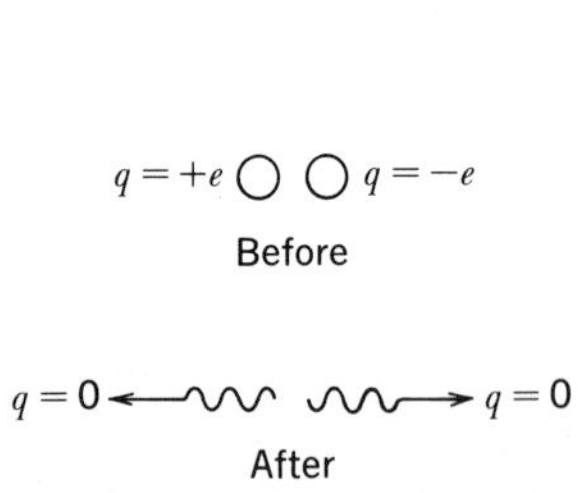

$q = +e \bigcirc \ \bigcirc q = -e$

Before

$q = 0 \longleftarrow\!\sim\!\sim \ \sim\!\sim\!\longrightarrow q = 0$

After

FIGURE 13-6   Two particles (an electron and a positron) having electric charge equal in magnitude but opposite in sign combine to form two new uncharged particles (gamma-ray photons).

Charge creation takes place in the inverse process, called *pair production.* Here a neutral gamma ray disappears and in its stead appear an electron and a positron. The analog to this process in momentum conservation is an explosion in which individual particles of the system attain momentum even though all were at rest initially. More about this in Section 28-2.

In later chapters we shall discuss in detail the fundamental particles of nature. Suffice it here to say that all known fundamental particles have just one of three possible charges: $+e$ ($= 1.6 \times 10^{-19}$ C), 0, or $-e$. Since any ordinary object is merely a collection of fundamental particles, the total charge of any charged object must be an *integral* multiple, positive or negative, of e. Electric charge is therefore *quantized;* it does not appear in arbitrary amounts but only in integral multiples of a smallest amount, or *quantum.* It has recently been speculated that the true quantum of charge might not be e but $\frac{1}{3}e$, but the *quark,* the elusive particle supposed to carry this charge, has not had its existence definitively established.

The total amount of charge on an ordinary charged object is so large compared to the basic quantum that for all practical purposes we can ignore its actual "graininess." This is like ignoring the graininess of sand when ordering 1 ton of sand. It is highly unlikely that the weight of a whole number of grains will amount precisely to 1 ton; but who cares? We could weigh the load to a precision of 1 part in 1 million, and still this error would correspond to hundreds of grains of sand.

Electric-charge quantization is present, under a different guise, in chemistry. The atomic number of any element, the integer that gives the element's place in the periodic table, is nothing more than the number of electrons in an electrically neutral atom. Quantization appears too in the concept of chemical valence. The valence, or combining power, of an element is measured by the excess charge on a single ion. Since this charge is quantized, the valence of any element is always exactly a positive or negative integer.

13-3 ELECTRIC-CHARGE QUANTIZATION

Electric-charge quantization: the charge of all objects is an integral multiple of the electron charge.

13-4 COULOMB'S LAW OF ELECTRIC INTERACTION

Much as the electric and gravitational interactions may differ in regard to charge, the forces of the two interactions obey essentially the same mathematical law. Specifically, if we (1) limit ourselves to a pair of charged *particles* q_1 and q_2, that is, objects whose size is negligible with respect to the distance separating them, and (2) suppose that these particles are stationary, we can, simply by the definition [13-1] of charge magnitude, assert that the magnitude of the force of q_2 on q_1, or $F_{2 \text{ on } 1}$ is proportional to q_1

$$F_{2 \text{ on } 1} \propto q_1$$

Similarly, the magnitude of the force of q_1 on q_2, or $F_{1 \text{ on } 2}$, is proportional to q_2

$$F_{1 \text{ on } 2} \propto q_2$$

But by Newton's third law the magnitudes of $F_{2\,\text{on}\,1}$ and $F_{1\,\text{on}\,2}$ are the same, and we designate the electric force on either q_1 or q_2 by F_e, where

$$F_e \propto q_1 q_2$$

There remains the question how the electric force between a pair of point charges varies with the distance r by which they are separated. We find by experiment that the electric force between any pair of point charges varies inversely as the square of the distance r, or

$$F_e \propto \frac{1}{r^2}$$

Therefore,

$$F_e \propto \frac{q_1 q_2}{r^2}$$

or, introducing a proportionality constant k_e,

[13-2]　$F_e = k_e \frac{q_1 q_2}{r^2}$

Coulomb's law for the electric interaction

Equation [13-2] describes the magnitude of the electric force. The line of the force is radial, i.e., along the line through the two charges. The direction (sense) along this line is determined by the relative signs of the two charges: for unlike charges, the attractive force (like the gravitational force) is directed toward the other particle; while for like charges the repulsive force is directed away from the other particle.

Coulomb's law [13-2] of electric interaction should be compared with Newton's law of universal gravitation [12-8],

$$F_g = G \frac{m_1 m_2}{r^2}$$

The two laws are seen to be identical in form. Moreover, condition 1 above, concerning the size of the particles, was also necessary for the law of gravitation. Condition 2, concerning their motion, however, was not; indeed Newton inferred his law from the *motions* of the planets. When one thinks about it, there is something strange about this second condition. It would seem to make Coulomb's law of very limited application; as soon as one or the other of the charges started moving under the influence of the force, the law would cease to hold! Actually, the condition can be relaxed. Experiment shows that [13-2] correctly describes the interaction force between two charged particles, *provided both of them are not in motion.* This means that Coulomb's law can be used to determine the force exerted on a moving charge q_1 by a stationary charge q_2; hence it determines the force on a charge moving in the presence of a number of stationary charges. But aren't things even stranger than before? What difference could it make in the force whether one particle moved and the other stayed fixed, or both moved, so long as the *relative* motion was the same in the two cases? This is a pregnant question. We shall start on the answer when we take up magnetism in Chapter 15, but the crux of the matter will not be reached until Chapter 21, with the theory of relativity. For now, let us explore the simpler consequences of [13-2].

The outstanding fact is that the electric force, as described by Coulomb's law, is a *central, inverse-square force.* Anything shown true of the gravitational force, in virture of its having those two properties, automatically holds for the electric force too. Thus, we have at once that *a spherical or spherically symmetric distribution of electric charge affects external bodies as though all the charge were concentrated at the center.* In fact, Charles Coulomb (1736–1806) first obtained [13-2] for two *finite* spheres (charged pith balls). His experiment (Fig. 13-7), though independent of Cavendish's, employed exactly the same sort of torsion balance, and we shall not discuss it further. The numerical value of the force constant is measured as

Spherical charge distribution acts on an external point charge as if all its charge were at the center.

$k_e = 8.98 \times 10^9$ N-m^2/C^2

$\approx 9 \times 10^9$ N-m^2/C^2 [13-3]

Compare the strengths of the electric and gravitational forces by computing the ratio of the electric force to the gravitational force between an electron (mass $= 9.1 \times 10^{-31}$ kgm) and a proton (mass $= 1.7 \times 10^{-27}$ kgm), both particles having a charge magnitude of 1.6×10^{-19} C.

EXAMPLE 13-4

From [13-2] and [12-8] we have

SOLUTION

$$\frac{F_e}{F_g} = \frac{k_e(q_1 q_2 / r^2)}{G(m_1 m_2 / r^2)} = \frac{k_e q_1 q_2}{G m_1 m_2}$$

$$= \frac{(9 \times 10^9 \text{ N-m}^2/\text{C}^2)(1.6 \times 10^{-19} \text{ C})^2}{(6.7 \times 10^{-11} \text{ N-m}^2/\text{kgm}^2)(9.1 \times 10^{-31} \text{ kgm})(1.7 \times 10^{-27} \text{ kgm})}$$

$$= 2.3 \times 10^{39}$$

Query To appreciate the enormous size of the number 10^{39}, carry out the following computation. Suppose that a very fast counter can count the number of atoms in a thimble full of water (about 10^{22} atoms) in a period of only 1 s. How long would it take for such a fast counter to reach 10^{39}? (The age of the universe is about 10^{17} s.)

FIGURE 13-7 Apparatus used by Coulomb in 1784 to establish that the repulsive force between two small spheres (*a* and *t*) charged with the same type of electricity is inversely proportional to the square of the distance between the centers of the spheres. A similar type of torsion balance was used later (1798) by Cavendish in determining the density of the Earth (Fig. 12-13).

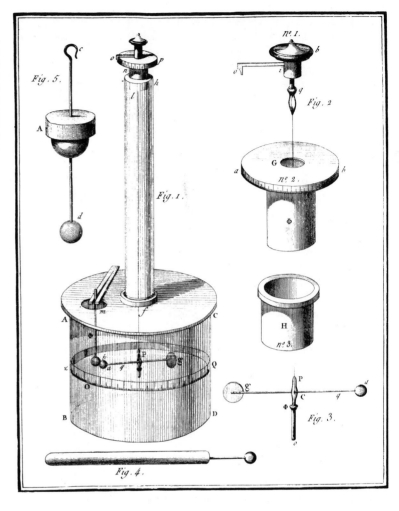

The electric attraction exceeds the gravitational attraction by a factor 2.3×10^{39} for all separation distances.

EXAMPLE 13-5   What is the net charge (spherically symmetric) on the Van de Graaff sphere in Examples 13-1 and 13-2.

SOLUTION   The electric force on a charge $q_1 = 10^{-10}$ C placed a distance of 1.0 m from the center of the sphere was 1.8×10^{-6} N. Therefore

$$F_e = \frac{k_e q_1 q_2}{r^2}$$

$$1.8 \times 10^{-6} = (9.0 \times 10^9)\frac{10^{-10} q_2}{1.0^2}$$

or

$$q_2 = 2.0 \times 10^{-6} \text{ C}$$

Three point charges are fixed at the positions shown in Fig. 13-8. Find the net force on q_2.

EXAMPLE 13-6

Since charges q_1 and q_2 have opposite signs, they attract each other; the force $\mathbf{F}_{1 \text{ on } 2}$ is to the left. Likewise q_2 and q_3 have opposite signs and attract each other; therefore, in this case the force $\mathbf{F}_{3 \text{ on } 2}$ is to the right. Thus the net electric force on q_2 is

SOLUTION

$$\mathbf{F} = \mathbf{F}_{1 \text{ on } 2} + \mathbf{F}_{3 \text{ on } 2}$$

$$= -k_e\frac{q_1 q_2}{r_{12}^2} + k_e\frac{q_2 q_3}{r_{23}^2}$$

$$= (+9 \times 10^9)\left[-\frac{(7 \times 10^{-8})(2 \times 10^{-8})}{9 \times 10^{-4}} + \frac{(2 \times 10^{-8})(8 \times 10^{-8})}{16 \times 10^{-4}}\right]$$

$$= -5.0 \times 10^3 \text{ N}$$

The net force is negative, i.e., to the left.

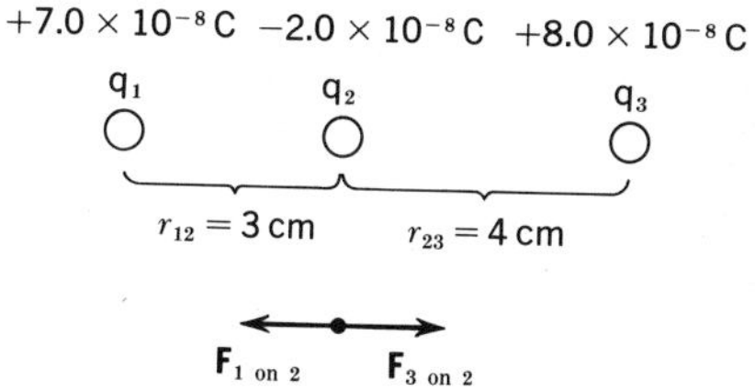

$+7.0 \times 10^{-8}\,\text{C} \quad -2.0 \times 10^{-8}\,\text{C} \quad +8.0 \times 10^{-8}\,\text{C}$

$q_1 \qquad q_2 \qquad q_3$

$r_{12} = 3\,\text{cm} \qquad r_{23} = 4\,\text{cm}$

$\mathbf{F}_{1 \text{ on } 2} \qquad \mathbf{F}_{3 \text{ on } 2}$

FIGURE 13-8

Brief mention was made in Chapter 12 of the notion of a field. We now return to this concept and in connection with the electric interaction develop it in detail.

13-5 THE ELECTRIC FIELD

Suppose that a single point charge q is placed near other fixed point charges. The resultant electric force on it can be found by applying Coulomb's law to its interaction with each of the other point charges and summing the results vectorially. Now imagine that this charge is replaced at the same location by a second charge $2q$. Without repeating the computation we know that the force on the new charge has the same direction as before but is twice as great. Again it is supposed that the nearby charges remain fixed in position. More generally, the resultant electric force on a charge at any one point is proportional to the magnitude of the charge, so that if the force on a unit charge is known, it is a simple matter to find the force on any other charge at the same location. This leads us to make the following definition. At each point of space the electric field **E** produced by one or more charged particles located in *fixed* positions is the electric force per unit charge experienced by a *positively* charge

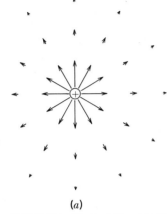

(a)

(b)

FIGURE 13-9 Electric field vectors at
selected positions in the vicinity of a
particle having (a) positive charge and
(b) negative charge.

particle placed at the point in question. Thus the electric field is a
vector which may vary from point to point. By placing a positive
test charge, or imagining such a charge to be placed, successively
at all points in the vicinity of the source charges, we can completely
ascertain the electric field.

Electric field **E**, the electric force per unit
charge

Since the electric field **E** is defined as the electric force per unit
charge, we can write

[13-4] $E = \dfrac{(F_e)_t}{q_t}$

E being the magnitude of the field at that point where the electric
force on the positive *test* charge q_t is $(F_e)_t$.

The simplest electric field is naturally that due to a single point
source q. If q is positive, it will repel a positive test charge, so that
the field vector **E** points everywhere radially outward (Fig. 13-9a).
If q is negative, **E** is directed radially inward (Fig. 13-9b). As for the
magnitude of **E**, it is by definition the magnitude of the force **F**$_t$ on
a test charge q_t divided by q_t. By Coulomb's law [13-2]

$$(F_e)_t = k_e \frac{q q_t}{r^2}$$

Hence, the magnitude of the electric field due to a point charge is

[13-5] $E = \dfrac{(F_e)_t}{q_t} = k_e \dfrac{q}{r^2}$

¹Or at least a very dense distribution. As we know,
electric charge is ultimately quantized.

The more elaborate field produced by a number of charged particles,
or even by a continuous distribution[1] of charge, can be found by

superposition. That is, the overall vector **E** can be obtained as the vector sum of the fields that would exist if the source charges or charge elements were present one at a time. The magnitude of each of these individual fields is given by [13-5].

Charges q$'$ = -6.4×10^{-10} C and q$''$ = $+3.6 \times 10^{-10}$ C are fixed at opposite corners of a rectangle 6 by 8 cm as shown in Fig. 13-10. What are the magnitude and direction of the electric field (*a*) at the center of the rectangle and (*b*) at the upper right corner?

EXAMPLE 13-7

a   The electric field **E** is simply the force which a unit positive charge would experience. Thus at the center the magnitude of the field produced by q$'$ is by [13-5]

SOLUTION

$$E' = \frac{(9 \times 10^9)(6.4 \times 10^{-10})}{(5 \times 10^{-2})^2} = 2.3 \times 10^3 \text{ N/C}$$

and the magnitude of the field produced by q$''$ is

$$E'' = \frac{(9 \times 10^9)(3.6 \times 10^{-10})}{(5 \times 10^{-2})^2} = 1.3 \times 10^3 \text{ N/C}$$

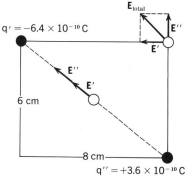

FIGURE 13-10

Because the two charges q$'$ and q$''$ have opposite signs, one force is attractive and the other repulsive. But the two charges are on opposite sides of the center; thus their fields are in the same direction. As a result the total field is 3.6×10^3 N/C along the diagonal of the rectangle pointing toward the negative charge q$'$.

b   At the upper right corner the field from the negative charge q$'$ is to the left with magnitude

$$E' = \frac{(9 \times 10^9)(6.4 \times 10^{-10})}{(8 \times 10^{-2})^2} = 0.9 \times 10^3 \text{ N/C}$$

and the field from the positive charge q$''$ is upward with magnitude

$$E'' = \frac{(9 \times 10^9)(3.6 \times 10^{-10})}{(6 \times 10^{-2})^2} = 0.9 \times 10^3 \text{ N/C}$$

Because both these contributions to the fields have the same magnitude, the total field simply makes a 45° angle above the top 8-cm side of the rectangle. The magnitude of this total field is $(\sqrt{2})(0.9 \times 10^{-3}) = 1.27 \times 10^3$ N/C.

To show completely in a single spatial diagram both the magnitude and direction of an electric field, one could, in principle, draw the

Electric field lines: their direction and density give the direction and magnitude of the electric field.

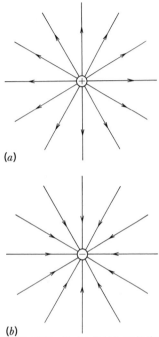

(a)

(b)

FIGURE 13-11   Electric field lines in the vicinity of a particle having (a) positive charge and (b) negative charge.

Electric field lines end on charges.

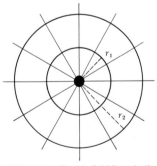

FIGURE 13-12   Electric field lines in the vicinity of a charged particle. The same number of lines, though more dispersed, cross a sphere of radius r_2 as cross the smaller sphere of radius r_1.

vector **E** at all points, but the result would be a jumble of arrows. Instead, use is made of *electric field lines*. An electric field line is a curve so drawn that the tangent at every point gives the direction of **E** at that point. Look at Fig. 13-11. We can no more display *all* the field lines than we could all the field vectors. We therefore agree to space the field lines according to the magnitude of **E.** More precisely, the number of field lines crossing a unit area perpendicular to **E** at any point will be made proportional to the magnitude of **E** at that point. Thus, the field lines will be densest where the field is strongest and sparsest where it is weakest. For any distribution of charges the field lines may always be constructed as *continuous* curves. It is never necessary, in order to make the number of lines per unit area proportional to the electric field magnitude, to end some field lines abruptly in empty space. This fortunate circumstance arises, basically, from the fact that the electric force between point charges is proportional exactly to the inverse *square* of the separation distance.

Consider a charged particle surrounded by two concentric spherical surfaces, as in Fig. 13-12. The smaller surface has radius r_1 and area $4\pi r_1^2$; the larger has radius r_2 and area $4\pi r_2^2$. By symmetry the electric field lines are rays from the source point, and by our convention their density in any one direction is the same as in any other. If, then, a fixed number of these lines is drawn from the source, the number of lines *per unit area* will decrease by the factor $(r_1/r_2)^2$ as one moves from the smaller to the larger spherical surface. This is precisely the factor by which **E** decreases, according to Coulomb's law. Hence no lines get lost or spring up between the two spheres. In general, for charges at rest *electric field lines begin and end only on charges*.

Back now to the meaning of the electric field. Since **E** at a given location is the force exerted there on a unit charge, a particle of charge q experiences a force

$$\mathbf{F} = q\mathbf{E} \qquad [13\text{-}6]$$

In this equation it is useful for q to be an algebraic quantity. Thus the implicit algebraic sign of q can denote the sign of the charge. Equation [13-6] then correctly describes the force—in the direction of **E** on a positive charge and in the opposite direction on a negative charge. We might say that [13-6] describes how charge q responds to the presence of an electric field. Such language would imply that the electric field is a real physical entity, one that exists quite apart from sources that produce it. Is this actually the case?

279

+ SECTION 13-6
THE ELECTRIC FIELD
OF A CONTINUOUS
DISTRIBUTION
OF CHARGE

For *static* fields (originating from charge distributions which are *fixed* or at least have been fixed for a long time) there is really no basis for asserting that the electric field is anything more than a useful artifice for simplifying certain computations. After all, nothing was arrived at in this section through the electric-field concept which could not have been derived directly from Coulomb's law.

But suppose the source charges are suddenly removed . . . .

Then, as Maxwell first showed in the nineteenth century, it is a different story. When charges shift position, a distant test particle is *not* affected at once. For instance, 1.28 s must pass before a charge on the moon reacts to the displacement of a charge on Earth (see Fig. 13-13). During that 1.28 s something—some thing—remains the same: the electric field near the lunar charge.

Maxwell, along with Faraday (who originated the idea of lines of force), conceived of the field as a disturbance in a nebulous medium filling all space, a medium sometimes referred to as the "ether". They thought of the medium between two points as being like a stretched rope, not only capable of transmitting a steady tension but also capable of transmitting bumpy disturbances that travel from one point to the other.

A revolution in the field concept occurred when Einstein developed the special theory of relativity. We now regard the electric field as having an autonomous existence, independent of its source and independent of any impalpable medium. We shall discuss these topics in later chapters; suffice it now to say that a wide range of physical experience testifies to the reality of electric fields.

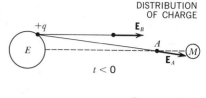

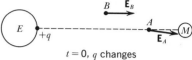

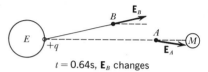

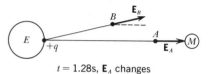

FIGURE 13-13 When the charge on an object, e.g., the antenna of a radio transmitter, located on the Earth is shifted, the electric field at points A on or near the moon does not change until sometime later. At points B about halfway to the moon it is necessary to wait only half as long to observe a change in the electric field.

Electric fields have a separate existence.

+ 13.6 THE ELECTRIC FIELD OF A CONTINUOUS DISTRIBUTION OF CHARGE

In many practical problems we want to find the motion of a charged particle in the presence of a large number of charged particles continuously distributed along a curve, over a surface, or throughout a volume. For example, in some television picture tubes the image is controlled by having electrons pass between two parallel sheets of charge, as shown in Fig. 13-14. A very large number of particles, each with a net negative charge, is distributed uniformly over the bottom of the upper plate, and the lower plate has a similar distribution of positive charge over its top surface. The magnitude of the total charge on each plate is the same, Q. To predict where an electron will strike the face of the tube we must know in detail the influence of the charged plates on its motion. Since the electron is attracted by the positive charge of the lower plate and repelled by the negative

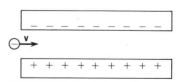

FIGURE 13-14 An electron with velocity **v** entering the region between two identical parallel plates having charges of equal magnitude but opposite sign.

charge of the upper, we know qualitatively that the electron will be deflected downward. It is, of course, possible to use Coulomb's law directly to determine the interaction force between the charged plates and some one electron entering the region between them at a particular position, but instead of specializing in one particular electron, it is more helpful to determine the electric field **E** at all points between the plates. Knowing this electric field, we can determine the force $\mathbf{F}_e = q\mathbf{E}$ on *any* particle introduced *anywhere* between the plates.

To calculate the field by applying Coulomb's law directly to each particle on the plates is laborious. An easier way is first to consider the electric field between two concentric spherical shells having the same separation distance d as the parallel plates. (see Fig. 13-15). The radius r_0 of the inner shell is chosen much larger than d. Moreover, we assume that positively charged particles are uniformly distributed over the inner shell so as to make the net charge *per unit area* the same as on the lower plate in Fig. 13-14. The total area A of the shell differs from the area a of the plate; consequently the total charge $+Q'$ on the spherical shell is different from the total charge $+Q$ on the lower plate. If the charge per unit area is designated $+\sigma$, we have

$$Q = \sigma a \qquad Q' = \sigma A = \sigma 4\pi r_0^2$$

The upper and lower plates are alike in magnitude of the total charge and area; therefore, the charge per unit area on the upper plate is just $-\sigma$. Let us also imagine that there is just enough negative charge on the outer shell in Fig. 13-15 for the charge per unit area on this shell also to be $-\sigma$. As a result the small segment of the shells circled in Fig. 13-15a and enlarged in Fig. 13-15b looks much like our set of parallel plates in Fig. 13-14. In fact, by taking r_0 sufficiently large with respect to d, we can ensure that surfaces of area a on the shells will be as flat as we please. Thus, when $r_0/d \gg 1$, the electric field between the plates may be assumed to be the same as that between the spherical shells, except at points near the edges of the plates. For such points the correspondence breaks down: there is no charge beyond the edges of the flat plates, but there is additional charge beyond the boundaries of the (essentially flat) spherical segments. However, this difference, which arises from the finiteness of the plates, is negligible (at least well in from the edges) if the dimensions of the plates are themselves large compared to d (they are) and if there is no large amount of *net* charge beyond the boundaries of the spherical segment (there is none).

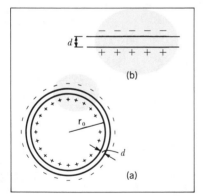

FIGURE 13-15 The parallel plates of Fig. 13-14 can be thought of as a small section of two large but closely spaced concentric spheres.

281

+ SECTION 13-6
THE ELECTRIC FIELD
OF A CONTINUOUS
DISTRIBUTION
OF CHARGE

Neglecting, then, any edge effects, we have reduced our problem to finding the electric field between the two spherical shells. This is easy to do. If a unit test charge is placed a distance r from the common center ($r_0 < r < r_0 + d$), we know from Section 13-4 that the inner shell acts on it like a particle of charge Q' located at the center. Moreover, the outer shell exerts no force at all on the test charge (again, just as in the case of gravity, and for the same reasons). Therefore the net force on the unit charge, which is the electric field, has the magnitude

$$E = \frac{k_e Q'}{r^2} = \frac{k_e 4\pi r_0^2 \sigma}{r^2} \approx 4\pi k_e \sigma$$

We can properly equate r to r_0 because d is so much smaller than r_0. The direction of **E** is radially outward because our positive test charge is repelled by the positive charge on the inner shell. Thus we conclude that the electric field lines in Fig. 13-15b are not only perpendicular to the small segments of the shells but very nearly parallel to each other.

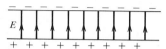

FIGURE 13-16 The electric field lines in the region between the charged parallel plates of Fig. 13-14.

But the electric field well in from the edges of the plates in Fig. 13-14 is like the *uniform* electric field in Fig. 13-15b. It has the magnitude

Electric field between oppositely charged parallel plates is uniform.

$$E = 4\pi k_e \sigma \qquad\qquad\qquad [13\text{-}7]$$

and is directed from the positive to the negative plate (Fig. 13-16). In this region an electron behaves just like a projectile near the Earth's surface. The positive plate acts as terra firma. In any calculations the actual weight of the electron may be ignored if, as usually is the case, the electric force is enormously larger.

A *solid sphere* of negative charge has a uniform density $\rho = 6.0 \times 10^{-6}$ C/m³. The radius of the sphere is $r_0 = 0.2$ m. What is the electric field at a distance of 3.0 m from the center of the sphere?

EXAMPLE 13-8

Because the charge distribution is spherically symmetric, the electric field outside the sphere is the same as though all the charge were at the center of the sphere. The total charge is the density times the volume

SOLUTION

$$Q = \rho \tfrac{4}{3}\pi r_0^3 = \frac{(6.0 \times 10^{-6})(4)(3.14)(8 \times 10^{-3})}{3} = 2.0 \times 10^{-7} \text{ C}$$

Therefore the magnitude of the electric field is

$$E = \frac{k_e Q}{r^2} = \frac{(9 \times 10^9)(2 \times 10^{-7})}{9} = 200 \text{ N/C}$$

Because the charge is negative, a positive test charge would be attracted toward the sphere. Thus the direction of **E** is radially inward.

EXAMPLE 13-9   Two concentric *spherical shells* of radii r_1 and r_2 ($r_2 > r_1$), shown in Fig. 13-17, have uniformly distributed negative charges $-q_1$ and $-q_2$, respectively. Find the electric field for the three regions (*a*) $0 < r < r_1$, (*b*) $r_1 < r < r_2$, and (*c*) $r > r_2$.

SOLUTION   *a*   For the interior region ($r < r_1$), the electric field is zero. Here a test charge is inside both shells of charge and accordingly experiences no net force.

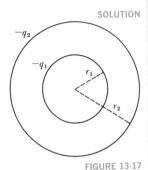

b   For the central region ($r_1 < r < r_2$), the test charge is inside the outer shell and experiences no net force from it. But the charge does experience an attractive (or inward) force from the negatively charged inner shell

$$E = k_e \frac{q_1}{r^2} \qquad \text{where } r_1 < r < r_2$$

FIGURE 13-17   *c*   For the exterior region ($r > r_2$), the test charge experiences attractive (or inward) forces from both of the negatively charged shells

$$E = k_e \frac{q_1 + q_2}{r^2} \qquad \text{where } r > r_2$$

EXAMPLE 13-10   Two large parallel plates are separated by a distance of 0.2 cm and charged to a density $+\sigma = +3.0 \times 10^{-6}$ C/m^2 on one plate and $-\sigma = -3.0 \times 10^{-6}$ C/m^2 on the other. What is the magnitude of the electric field between the plates (see Fig. 13-16)?

SOLUTION   For a set of parallel plates where the edge effects can be neglected,

$$E = 4\pi k_e \sigma = 4(3.14)(9 \times 10^9)(3 \times 10^{-6}) = 3.4 \times 10^5 \text{ N/C}$$

EXAMPLE 13-11   A particle of mass m and positive charge q enters a region where there is a uniform downward electric field **E**. The initial velocity of the particle is horizontal, and its initial speed is v_0 (see Fig. 13-18). What is the vertical deflection y_L of the particle after it has moved a horizontal distance L through the uniform field?

SOLUTION   Compare Example 5-5. No horizontal force acts on the particle, and it coasts horizontally at constant speed. If the particle covers the horizontal distance L in a time Δt, then $\Delta t = L/v_0$. The vertical speed will change with constant acceleration because of the constant vertical force

283

+   SECTION 13-6
THE ELECTRIC FIELD
OF A CONTINUOUS
DISTRIBUTION
OF CHARGE

$$F_y = \frac{\Delta m v_y}{\Delta t}$$

or

$$qE = \frac{m v_y - 0}{L/v_0}$$

Therefore

$$v_y = \frac{qEL}{m v_0}$$

and

$$y_L = \langle v_y \rangle \, \Delta t = \frac{0 + v_y}{2} \frac{L}{v_0} = \frac{qEL^2}{2 m v_0{}^2}$$

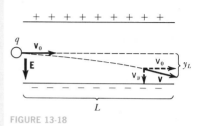

FIGURE 13-18

EXAMPLE 13-12

Show directly from Coulomb's law that the electric field *near* the surface of a large uniform plane sheet of charge has (*a*) a direction perpendicular to the sheet of charge and (*b*) a magnitude that is the same everywhere.

a   We can show immediately, simply on the basis of symmetry, that the electric field must be perpendicular to the plane. For suppose that the electric field **E** were not perpendicular to the plane. Then there would exist a component parallel to the surface pointing in some one direction. This, in turn, would imply that the direction along the plane is preferred over any other direction. But this is impossible, because we assumed the plane to be uniformly charged, in which case no one direction is preferred over any other. It follows, then, that the electric field can have no component along the plane; it must be perpendicular to the plane.

b   To find how the magnitude of the field at a point P depends on its distance h above the plane we imagine the sheet to be subdivided into small squares, each of area A, as shown in Fig. 13-19, and concentrate on the electric field produced by the square shown shaded. The area A is so small that the charge on it is approximately equivalent to a single point charge, and it produces an electric field **E** at P. We could imagine point P of Fig. 13-19 to be shifted upward to increase h; equivalently, and more simply for the present analysis, we suppose P to be fixed while the sheet is moved downward, so that P is now a distance h' from the sheet. The checkerboard pattern on the lower sheet is such that corresponding corners of small square of area A and the larger square of area A' lie on straight lines passing through P, or a small area A of the upper sheet is projected by an imaginary searchlight

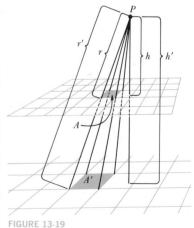

FIGURE 13-19

beam from P into the area A' on the lower sheet. From the geometry of Fig. 13-19 we see that the sides of the larger and smaller squares are in the same ratio as h' and h, so that the area ratio is $A'/A = (h'/h)^2$. Similarly, the distances r and r' of the two squares from P are in the ratio $r'/r = h'/h$. The electric force on a unit test charge at P from the charge in the area A is proportional to the amount of charge in A, and therefore proportional to A, and the force, also from Coulomb's law, is inversely proportional to r^2. How does this force compare with that arising from the charge on the larger area A' at a larger distance r' from P? The amount of charge is larger, since the area A' is larger, by the factor $A'/A = (h'/h)^2$; but the distance r' is also larger by the factor $r'/r = h'/h$, and the distance squared is larger by the factor $(h'/h)^2$. The two factors just cancel, and the electric field at P from area A' at distance r' is exactly the same as the electric field from area A at r. If the electric field from any one square does not depend on the distance h of P from the sheet, the electric field due to all the squares on the sheet must be independent of h. We have proved that the magnitude of the electric field from a uniform sheet is the same everywhere (provided only that the distance from the sheet is small compared to the size of the sheet so that any effects from near the edges are negligible).

Does the pattern of this proof seem familiar? Look again at Example 12-4, which gives Newton's demonstration that the gravitational field is zero in a spherical shell of material. The whole story of inverse-square forces lies in this seesaw of area and distance.

Query What is the field below the sheet of charge in Example 13-12?

Query Given two large horizontal sheets, one of positive and the other of negative charge, separated by a distance d, what is the field above, between, and below these sheets (see [13-7])? Therefore, what is the magnitude of the electric field in Example 13-12?

Query The electric field from a uniformly charged spherical shell is radial and its magnitude falls off with the inverse square of the distance from the sphere's center. The electric field from a uniformly charged plane is perpendicular to the plane and constant in magnitude. Can you guess, using symmetry arguments and analogy, the character of the electric field from a long uniformly charged right-circular cylinder?

+ 13-7 ENERGY CONTAINED IN THE ELECTRIC FIELD

Earlier it was noted that under suitable circumstances an electric field can be made to travel as a separate physical entity carrying energy and momentum through otherwise empty space. Using the results of the previous section, we can now find the expression for the energy associated with an electric field.

Two large sheets of charge are shown in Fig. 13-20. The sheets are identical except for the sign of the constant charge density. Initially they are very close together, nearly touching, as in Fig. 13-20a. Then, with the lower plate fixed in position, the upper plate is slowly pulled out to a distance d (Fig. 13-20b). A uniform electric field exists between the plates, inasmuch as the separation d is taken to be small compared to the dimensions of the sheets. In magnitude this electric field is [13-7]

$$E = 4\pi k_e \sigma$$

irrespective of the value of d, so long as d is small.

Since the opposite charges on the two sheets attract each other, it is clear that work was required to separate them. The net charge on the upper sheet is $Q = \sigma A$, where A is the area of either sheet. Thus, the net electric force attracting the upper sheet to the lower sheet is $F = QE_1 = \sigma A E_1$, where E_1, the field from the lower sheet alone is $\frac{1}{2}E$. The work done in moving the sheets apart at constant speed is then

$$W = Fd = \sigma A E_1 d$$

Since work is done on the system but neither sheet acquires kinetic energy, the electric potential energy of the system must have increased by an amount $\sigma A E_1 d$. We can, as usual, identify this change in potential energy with a change in the system's configuration, i.e., with the change in separation of the opposite charges on the two sheets. But we can also identify the increased potential energy of the system with the electric field which has been created between the two plates by doing work on the system. Recognizing that the electric field exists throughout the volume $V = Ad$ between the two separated sheets, we can write

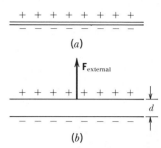

FIGURE 13-20 (a) Two large, closely spaced sheets having charges of equal magnitude but opposite sign. (b) While the lower sheet is held fixed, the upper sheet is raised by an externally applied upward force.

$$W = \sigma A E_1 d = \left(\frac{E}{4\pi k_e} A E_1 d \right) = \frac{E^2}{8\pi k_e} V = \text{electric potential energy}$$

where we have set $E_1 = \frac{1}{2}E$.

Then the *energy density* of the electric field, the energy per unit volume, is

The electric-field energy density is proportional to the square of the electric field.

$$\text{Energy density} = \frac{\text{energy}}{\text{volume}} = \frac{E^2}{8\pi k_e} \qquad [13\text{-}8]$$

Though derived here for the special case of a uniform field, [13-8] applies to any small volume of a general field, provided the local value of E is inserted. Note that the electric energy density is proportional to the *square* of the electric field. This implies that to double the electric field in the same volume of space takes 4 times as much work. In our example, this follows from the fact that one first doubles the charge on the sheets while they are close together; then, not only is the attractive force doubled, the amount of charge to be moved against that force is doubled. The two factors together increase the work by a factor of 4.

Electric charge (1) comes in two varieties, positive and negative, (2) SUMMARY is conserved for an isolated system, and (3) is quantized.

The electric force between point charges q_1 and q_2 separated by a distance r is radial and has a magnitude given by the Coulomb law relation

$$[13\text{-}2] \quad F_e = k_e \frac{q_1 q_2}{r^2}$$

where $k_e \approx 9 \times 10^9$ N-m^2/C^2. Charges of unlike sign attract, and like charges repel.

The electric field **E** is defined as the electric force per unit charge on a positive test object

$$E = \frac{(F_e)_t}{q_t}$$

A particle of charge q in an electric field **E** experiences a force

F = qE

An electric field line is a curve so drawn that the tangent at every point gives the direction of **E**.

The electric field from a point charge or outside a spherically symmetric charge distribution is everywhere radial. The electric field between two uniformly charged plane parallel sheets of opposite sign is everywhere uniform and perpendicular to the planes.

PROBLEMS

13-1   A standard charge of 10^{-10} C experiences an electric force of 3.5×10^{-6} N when placed near a large charged sphere. When the standard charge is removed and an object of unknown charge placed in the same position, the new object experiences an electric force of 1.75×10^{-5} N. What is the net electric charge on this new object?

13-2   (a) Calculate the magnitude of the force between two protons separated by a distance comparable to the diameter of a nucleus, i.e., about 5×10^{-15} m. (b) This force is comparable to the weight of what amount of mass at the Earth's surface? (c) Would you say that it takes weak or strong forces to keep these 10^{-27} kgm particles bound within the nucleus?

13-3   In 1 gm of hydrogen there are approximately 6.0×10^{23} atoms, each consisting of one proton and one electron. Imagine that all the electrons in this gram of hydrogen were somehow moved to the North Pole of the Earth and all the protons to the South Pole, about 1.3×10^7 m away. What would be the force of attraction between the electrons and protons?

13-4   Three point charges are arranged in a straight line as shown in the figure. Find both the magnitude and direction of the net force acting on q_2.

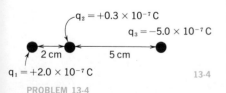

$q_2 = +0.3 \times 10^{-7}$ C

$q_3 = -5.0 \times 10^{-7}$ C

2 cm   5 cm

$q_1 = +2.0 \times 10^{-7}$ C

PROBLEM 13-4

A small pith ball with a mass of 4.0 gm and a charge $q_0 = -1.00 \times 10^{-8}$ C is suspended from an insulating thread which makes an angle of $37°$ with the vertical when two point charges are located near it, as shown in the figure. The point charge on the right has the value $q_1 = +1.0 \times 10^{-6}$ C. What are the magnitude and sign of the other point charge? **13-5**

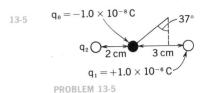

$q_0 = -1.0 \times 10^{-8}$ C

q_2 $\quad$ 2 cm $\quad$ 3 cm

$q_1 = +1.0 \times 10^{-6}$ C

PROBLEM 13-5

A point charge $+1.0 \times 10^{-7}$ C is placed 10 cm from a second point charge of -4.0×10^{-8} C. Where then must a third point charge of $+5 \times 10^{-8}$ C be placed if there is to be no net force on this third charge? **13-6**

Two point charges of equal magnitude exert a force of 2.0 N on each other. When their separation distance is decreased by 0.50 cm, the force increases to 18.0 N. (*a*) What was their initial separation? (*b*) What is the magnitude of each charge? **13-7**

A small object of mass 10 gm and charge -2.0×10^{-7} C is placed in a uniform electric field **E**. What must be the magnitude and direction of this field if the object is to remain stationary near the surface of the Earth under the combined influence of **E** and the Earth's gravitational attraction? **13-8**

Four charges are fixed at the corners of a rectangle measuring 6 by 8 cm, as shown in the figure. (*a*) What are the direction and magnitude of the net force on a -4×10^{-7} C charge at the center? (*b*) What are the direction and magnitude of the electric field at the center? **13-9**

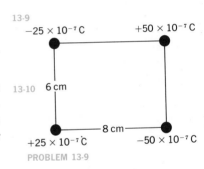

-25×10^{-7} C $\qquad$ $+50 \times 10^{-7}$ C

6 cm

$+25 \times 10^{-7}$ C $\quad$ —8 cm— $\quad$ -50×10^{-7} C

PROBLEM 13-9

There is evidence that the magnitude of the electric field just above the earth's surface averaged over all locations is not zero but about 10 N/C. This field is directed toward the center of the Earth. How many excess electrons per square meter on the surface of the Earth would account for this observation? **13-10**

In order to obtain a force of 10^{-2} N, that is, the weight of 1 gm, on a test object which has an electric charge of 10^{-10} C (*a*) what total charge would have to be placed uniformly on a sphere the center of which is 1 m from the test object or (*b*) what charge per unit area would have to be placed on two oppositely charged parallel plates between which the test object is located? **13-11**

The electric field from a *single* infinitely large plane sheet of charge of uniform density σ is everywhere perpendicular to the sheet and everywhere of magnitude $2\pi k_e \sigma$. By considering the parallel plates of [13-7] to be equivalent to two large plane sheets of charge, explain how [13-7], $E = 4\pi k_e \sigma$ for the plates, is consistent with $E = 2\pi k_e \sigma$ for a single sheet. **13-12**

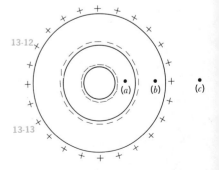

Three concentric spherical shells of charge -2.0×10^{-6}, -6.0×10^{-6}, and $+8.0 \times 10^{-6}$ C have radii of 2.0, 4.0, and 8.0 cm respectively. What are the magnitude and direction of the electric field at the radial distance (*a*) $r = 3.0$ cm, (*b*) $r = 6.0$ cm, and (*c*) $r = 10.0$ cm? **13-13**

PROBLEM 13-13

13-14 Why is it impossible to derive the expression for the electric field of a single plane sheet of charge ($E = 2\pi k_e \sigma$) directly from the known electric field of a large thin shell of electric charge ($E = k_e Q'/r_0^2$) (see Section 13-6)?

13-15 In Example 13-12 it was proved that the electric field from a uniformly charged infinite plane surface is perpendicular to the plane and constant in magnitude for all distances from the plane. The Coulomb force between point charges was taken, of course, to be inverse-*square*. Suppose, for the sake of argument, that the electric interaction between point charges were inverse-*cube*. What would then be (*a*) the direction and (*b*) the distance-dependence of the magnitude of the electric field from a uniformly charged plane surface?

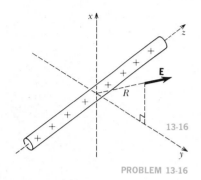

PROBLEM 13-16

13-16 A very long straight length of thin metal tubing is uniformly charged. Use symmetry arguments to show that the direction of the net electric field at any point outside the tube is perpendicular to the tube and directed along a radial line from the center of the tube.

c 13-17 The long straight thin wire shown in the figure is uniformly charged with λ C/m. (*a*) Show that the increment of charge $dq = \lambda\, dz$ contributes an increment to the radial electric field of magnitude $dE_R = (\lambda\, dz/s^2) \cos \beta$, where $s = R/(\cos \beta)$, $z = R \tan \beta$, and $dz = R \sec^2 \beta\, d\beta$. (*b*) Prove that the net electric field is directed radially outward and has magnitude $E_R = k_e(2\lambda/R)$.

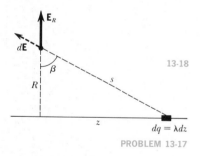

PROBLEM 13-17

13-18 Two electric charges of equal magnitude but opposite sign separated by a fixed distance comprise what is called an *electric dipole*. Consider a dipole, with charges $+q$ and $-q$ and separation d, centered at the origin and aligned along the y axis. (*a*) Show that the electric field at points far out on the y axis has the value

$$E(x = 0, y) \approx k_e \frac{2qd}{y^3} \qquad (y \gg d)$$

(*b*) Show that the electric field at points far out on the x axis has the value

$$E(x, y = 0) \approx k_e \frac{qd}{x^3} \qquad (x \gg d)$$

(*c*) What is the direction of **E** at these various locations on the x and y axes? (*d*) Draw approximate electric field lines showing the directions of **E** elsewhere.

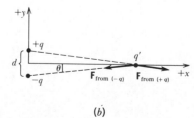

(*a*)

(*b*)

PROBLEM 13-18

MORE ON THE ELECTRIC INTERACTION

CHAPTER FOURTEEN

14-1 ENERGY CONSERVATION AND ELECTRIC POTENTIAL

Another important property of the electric interaction remains to be discussed. The energy-conservation principle can be used for any isolated system where the interaction force is *conservative,* i.e., for any isolated system where the total change in kinetic energy (work done by the interaction forces) is zero around any possible closed path followed by a particle. We have already seen in Section 12-7 that the gravitational interaction is conservative. But the electric force between charged particles, like the gravitational force, is radial, with a magnitude that depends only on the particle separation. Thus the electric interaction is also conservative, and we can define and calculate a change in electric potential energy for use in the energy-conservation principle.

The electric force is conservative, and a potential energy can be associated with it.

Let us calculate the change in electric potential energy for two simple situations, both analogous to situations we have already discussed for the gravitational interaction: (1) a particle interacting with the oppositely charged parallel plates of Fig. 14-1 (a constant interaction force) and (2) a particle interacting with a spherically symmetric charge distribution (a radial interaction force).

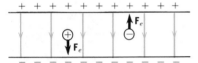

FIGURE 14-1 Direction of the electric force on a positively (negatively) charged particle between two uniformly charged parallel plates.

If a particle is shot in the direction *opposite* to the constant force on it between the parallel plates of Fig. 14-1, it will lose kinetic energy, just as a ball thrown straight up (the direction opposite to the force of gravity) loses kinetic energy. As the charged particle moves opposite to the electric force, the system gains electric potential energy, just as the Earth-ball system gains gravitational potential energy as the ball rises. Indeed, the calculation of the change in electric potential energy is just like the gravitational calculation of Section 9-3, only now the dominant interaction force is not the gravitational force mg but the much stronger electric force given by [13-6] and [13-7]

$$[14-1] \quad F_e = q(4\pi k_e \sigma)$$

Because our parallel plates are very massive compared to the charged particle, we assume, as we usually did for the gravitational calculations, that the total kinetic energy of the system is essentially the kinetic energy of the particle only. Thus using [8-7] we find that the change in kinetic energy of the system as the particle moves in the direction opposite to the electric force is

$$\Delta E_k = F_x \Delta x = -F_e \Delta x$$

where Δx is the displacement (just as Δh was the displacement, or change in height, in the gravitational calculation). Finally, from the energy-conservation principle, we have

$$[14-2] \quad \Delta E_p = -\Delta E_k = F_e \Delta x$$

291

SECTION 14-1
ENERGY
CONSERVATION
AND ELECTRIC
POTENTIAL

where the right-hand side is the exact analog of $mg \, \Delta h$; the magnitude of the electric force F_e from [14-1] has replaced the magnitude mg of the gravitational force, and the displacement Δx (in the direction opposite to the force) has replaced the change in height Δh. Equation [14-2] makes sense. When a particle moves opposite to the electric force ($\Delta x > 0$), the system loses kinetic energy but gains potential energy. And when the particle moves in the direction of the electric force ($\Delta x < 0$), the system gains kinetic energy but loses potential energy.

Equation [14-2] has been derived for the special case where the displacement is along a line parallel to the direction of the constant electric force. But by same argument used in Example 8-9, it follows that for a *constant* electric force [14-2] correctly describes the change in electric potential energy for *any* displacement: the change in electric potential energy depends only on the component Δx of the displacement parallel to the direction of the electric force.

In Section 13-5 we used the idea of the electric field **E** to describe the electric interaction of a charged particle placed in the vicinity of a fixed charge distribution. The electric field was defined as the electric force per unit charge exerted by the distribution on a *positive* test object. Correspondingly, it is often convenient to describe the energy changes by using a new quantity, the change in electric potential energy *per unit charge* of the positive test object. These potential-energy changes occur as the test object of charge q_t moves from one location to another. The new quantity, called the *change in electric potential* (or sometimes, electric potential difference) is given by

Electric potential change: change in electric potential energy per unit positive charge

$$\Delta V = \frac{(\Delta E_p)_t}{q_t}$$

[14-3]

The units of electric potential difference in the mks system are joules per coulomb, a combination used so frequently that it is assigned the shorter name *volt* (V).

Because the change in electric potential ΔV is the change in electric potential energy as a unit positive charge moves between two positions, the change in electric potential energy when *any* particle having charge q moves between these two positions is

$$\Delta E_p = q \, \Delta V$$

[14-4]

Here, as in [13-6], it is useful for q to be an algebraic quantity whose sign denotes the sign of the charged particle. Because ΔV is defined in terms of a positive test object, the sign of ΔE_p for a particle of positive charge q is the same as the sign of ΔV. But the sign of ΔE_p for a negatively charged particle must be opposite to the sign of ΔV.

Electron volt defined

We can use [14-4] to define a new unit of energy, the *electron volt* (eV), used frequently in atomic physics. One electron volt is the magnitude of the change in electric potential energy when an electron (charge $e = 1.6 \times 10^{-19}$ C) moves through an electric potential difference of 1 volt. Thus, using [14-4]

[14-5] $1.0 \text{ eV} = (1.6 \times 10^{-19} \text{ C})(1.0 \text{ V}) = 1.6 \times 10^{-19} \text{ J}$

(Energy conservation can be used to give an alternate statement of the meaning of 1 eV; see Example 14-3.)

For a test charge q_t moving between the parallel plates of Fig. 14-1, the change in electric *potential energy* is given by [14-2]. Therefore, the change in electric *potential,* defined by [14-3], is

$$\Delta V = \frac{(F_e)_t \, \Delta x}{q_t}$$

But using the definition of the electric field [13-4] this becomes

[14-6] $\Delta V = E \, \Delta x$

Recall that the positive x direction is opposite to the direction of the electric force (which for a positive charge is the direction of **E**). Thus the change in potential is positive for a displacement opposite to the electric field ($\Delta x > 0$) and negative for a displacement in the direction of the electric field ($\Delta x < 0$). Said differently, electric field lines point in the direction of decreasing electric potential. This is exactly what we expect; a positive charge, starting from rest, moves in the direction of the electric field, gains kinetic energy, and loses electric potential (and electric potential energy).

EXAMPLE 14-1

What is the electric potential difference between two parallel plates separated by 0.2 cm, if the uniform electric field between them is 3.4×10^5 N/C? Which is at the lower potential?

SOLUTION

The electric field is uniform between these plates. Therefore, using [14-6]

$\Delta V = E \, \Delta x = (3.4 \times 10^5 \text{ N/C})(2 \times 10^{-3} \text{ m}) = 680 \text{ J/C} = 680 \text{ V}$

The electric potential always decreases as one moves in the direction of the electric field lines because a positive charge moving freely in this direction gains kinetic energy but loses potential energy. There-

293

SECTION 14-1
ENERGY
CONSERVATION
AND ELECTRIC
POTENTIAL

fore, with the electric field lines extending *from* the positively charge plate *to* the negatively charged plate, the negatively charged plate is at the lower electric potential.

Solve Example 13-11 again, using the energy-conservation principle.

EXAMPLE 14-2

For the isolated system of parallel plates plus particle

$$\Delta E_k = -\Delta E_p$$

Since the plates are assumed to be very massive, all the kinetic-energy change is associated with the particle. Furthermore, the particle is assumed to have so small a mass and the electric field is assumed so large that the change in gravitational potential energy is negligible compared to the change in electric potential energy.

The electric force is downward, and so as the particle moves closer to the lower plate, it loses potential energy

$$\Delta E_p = F_e\,\Delta y = -qEy_L$$

The particle's final velocity **v**, with horizontal component v_0 and vertical component v_y, is given by $v^2 = v_0{}^2 + v_y{}^2$. Thus the energy-conservation principle can be written

$$\tfrac{1}{2}mv^2 - \tfrac{1}{2}mv_0{}^2 = -(-qEy_L)$$
$$\tfrac{1}{2}mv_y{}^2 = qEy_L$$

But for motion at constant acceleration (y component)

$$y_L = \tfrac{1}{2}v_y\,\Delta t$$

and for motion at constant velocity (x component)

$$L = v_0\,\Delta t$$

Therefore, using these last two equations, we also have

$$v_y{}^2 = \left(\frac{2y_L}{\Delta t}\right)^2 = \left(\frac{2y_Lv_0}{L}\right)^2$$

Substituting this into the energy relationship,

$$\tfrac{1}{2}m\left(\frac{2y_Lv_0}{L}\right)^2 = qEy_L$$

$$y_L = \frac{qEL^2}{2mv_0{}^2}$$

EXAMPLE 14-3

A ———————————— +20.0 V

B ———————————— 0 V

C ——— v_0↑ ——— +4.0 V

FIGURE 14-2

The electric potentials over the three surfaces A, B, and C, shown in Fig. 14-2, are $+20.0$, 0, and $+4.0$ V, respectively. An electron is projected upward from surface C with an initial kinetic energy of 9.0 eV. (Remember that 1 eV is the kinetic energy acquired by an electron as it goes through an electric potential difference of 1 V.) Surface B is porous, so that the electron can pass through it if it arrives there. (a) Will the electron, in fact, arrive at surface B, and if so, with what kinetic energy? (b) Will it get to surface A, and if so, with what kinetic energy?

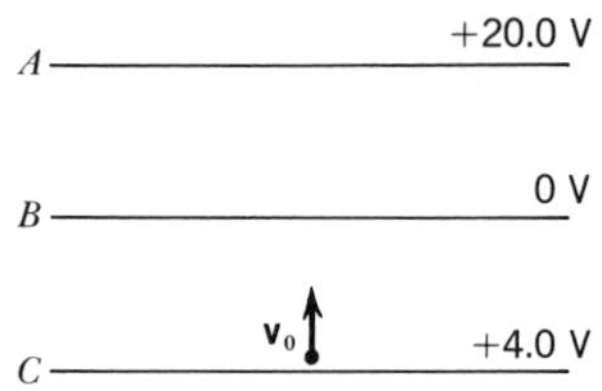

SOLUTION

a Inasmuch as the electric potential decreases from C to B, the electric field E in this region is directed upward from C to B. Consequently, the electric force $\mathbf{F}_e$ on the negatively charged electron will be directed downward, toward C, and the electron will be slowed. By energy conservation we know that it loses kinetic energy of 4.0 eV; thus, the electron will reach surface B with a remaining kinetic energy of $9.0 - 4.0 = 5.0$ eV.

b The electric potential difference is reversed in the region between B and A, and so the electron will be speeded up there. It will gain 20.0 eV of kinetic energy, arriving at surface A with a kinetic energy of $5.0 + 20.0 = 25$ eV.

+ 14-2 ELECTRIC POTENTIAL ENERGY FOR A RADIAL INTERACTION FORCE

For a spherically symmetric charge distribution, it is not necessary to go through a detailed calculation of the change in potential energy; we can obtain the result simply by looking at the analogous calculations for the gravitational interaction. In Section 12-7 we obtained expression [12-12] for the change in gravitational potential energy of an isolated system of two spherically symmetric bodies as their separation distance changed from an initial value r_i to a final value r_f

$$(\Delta E_p)_{\text{grav}} = Gm_1 m_2 \left(\frac{1}{r_i} - \frac{1}{r_f} \right)$$

Comparing [12-8] and [13-2], we see that the analog of $Gm_1 m_2$ is $k_e q_1 q_2$. Therefore, recalling that the gravitational interaction is attractive, we conclude that the change in electric potential energy for an isolated system of two spherically symmetric charges q_1 and q_2 which *attract* each other (charges of unlike sign) through the electric interaction is:

Electric potential energy for point charges

Attractive interaction (charges of unlike sign):

[14-7] $(\Delta E_p)_{\text{elec}} = k_e q_1 q_2 \left(\frac{1}{r_i} - \frac{1}{r_f} \right)$

295

+ SECTION 14-2
ELECTRIC POTENTIAL
ENERGY FOR A
RADIAL INTERACTION
FORCE

Moreover, if we arbitrarily assume that the electric potential energy is zero when the charges are essentially infinitely far apart (attractive force is zero), we can formulate the energy-conservation principle in terms of absolute values of potential energy. By comparison with the analogous result for the attractive gravitational interaction, [12-13], we conclude that the potential energy in this situation is

Attractive interaction (charges of unlike sign):

$$(E_p)_{elec} = \frac{-k_e q_1 q_2}{r} \qquad [14\text{-}8]$$

Of course, for the repulsive interaction of two charges (like sign), there is a decrease in potential energy as the charges get farther apart. Thus the expressions for ΔE_p and E_p are similar to [14-7] and [14-8], but they have the opposite sign:

Repulsive interaction (charges of like sign):

$$(\Delta E_p)_{elec} = -k_e q_1 q_2 \left(\frac{1}{r_i} - \frac{1}{r_f} \right) \qquad [14\text{-}9]$$

$$(E_p)_{elec} = \frac{+k_e q_1 q_2}{r} \qquad [14\text{-}10]$$

Finally, by identifying the charge q_2 with a positive test object having charge q_t and the charge q_1 with a spherically symmetric source distribution having net charge q, we can use our definition of electric potential, [14-3], and, for example, [14-9] and [14-10] to obtain expressions for both the *change* in electric potential and the electric potential in the vicinity of the source charge

Electric potential

$$\Delta V = -k_e q \left(\frac{1}{r_i} - \frac{1}{r_f} \right) \qquad [14\text{-}11]$$

$$V = \frac{+k_e q}{r} \qquad [14\text{-}12]$$

The signs in these equations make sense: if the source charge consists of positive charge, the interaction force with a positive test object is repulsive; then during an outward displacement of the test object $(r_i < r_f)$ the kinetic energy increases and the electric potential decreases, hence the minus sign in [14-11]. For a source of negative charge, the interaction force with the positive test object is attractive, and all the signs are reversed. We account for this simply by using the algebraic symbol for q in [14-11] and [14-12]; then the reversal of sign for a negative source comes automatically.

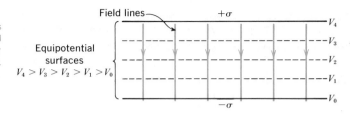

FIGURE 14-3 Plane equipotential surfaces everywhere perpendicular to the parallel electric field lines between two uniformly charged parallel plates.

+ 14-3 EQUIPOTENTIAL SURFACES

Surface of constant electric potential: equipotential surface

Electric field lines are perpendicular to equipotential surfaces.

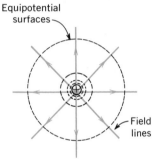

FIGURE 14-4 Spherical equipotential surfaces everywhere perpendicular to the radial electric field lines from a positive point charge.

Just as an electric field can be depicted graphically by electric field lines, electric potential can be represented by surfaces. A surface of constant electric potential, called an *equipotential surface,* is one over which a test charge can be moved without changing the electric potential energy. *Electric field lines and equipotential surfaces are mutually perpendicular.* Otherwise, at some point **E** would have a nonzero component along an equipotential surface. This component would do work on a test charge moving in the surface, thereby producing a potential-energy change. But by definition this is impossible. The equipotential surfaces are easily constructed when the field lines are known, and vice versa.

Figure 14-3 shows the electric field lines and the associated equipotential surfaces for the region between two oppositely charged parallel plates. Figure 14-4 shows the region around a positive point charge, and Fig. 14-5 applies to a spherical shell with negative charge uniformly distributed over its surface. Note that, in accordance with our earlier statement, the field lines always point in the direction of decreasing electric potential.

EXAMPLE 14-4

An electron $(-q_1 = -1.6 \times 10^{-19}$ C) and a positron $(+q_2 = +1.6 \times 10^{-19}$ C) start from rest 1.0 m apart. As a result of their electric interaction they accelerate toward each other. What is the kinetic energy in electron volts of each when they are 1 angstrom (Å) apart? (Both particles have the same mass, 9.1×10^{-31} kgm, and 1.0 Å $= 10^{-10}$ m, about the diameter of an atom.)

SOLUTION

According to the energy-conservation principle, the change in kinetic energy of the system is equal in magnitude but opposite in sign to the system's change in electric potential energy given by [14-7] (attractive interaction)

$$\Delta E_k = -\Delta E_p$$
$$2(\tfrac{1}{2}mv^2 - 0) = -k_e q^2 \left(\frac{1}{r_i} - \frac{1}{r_f} \right)$$

where the factor of 2 on the left side accounts for the fact there are *two* particles of equal mass starting from rest and attaining the same final speed v. Therefore, the final kinetic energy of *each* particle is

$$\tfrac{1}{2}mv^2 = -\tfrac{1}{2}(9 \times 10^9)(1.6 \times 10^{-19})^2(1 - 10^{10})$$

$$\approx \frac{11.5 \times 10^{-19} \text{ J}}{1.6 \times 10^{-19} \text{ J/eV}} = 7.2 \text{ eV}$$

The total kinetic energy of the system is 14.4 eV when the two particles are 1 Å apart.

Query What would the answer in Example 14-4 be if the particles had started 1,000 m instead of 1.0 m apart?

When a charged metal sphere is touched by materials like glass, rubber, string, wax, oil, or cloth, measurement shows that the sphere retains its charge. Materials such as these are called *insulators*. On the other hand, when touched by your hand, a piece of carbon, the ground, or almost any damp material, the sphere readily loses its charge. Materials which can lead away charge are called *conductors*. Most metals are excellent conductors.

Charge is transferred through metallic conductors by the shifting of electrons in the metal. In most materials electrons are rather tightly bound to their parent nuclei, but in metals about one electron per atom is not held tightly. Such a free electron can move through the conductor with relative ease.

Two metal plates are shown in Fig. 14-6a. The right plate is negatively charged, with an excess of electrons; the one on the left is positively charged, with a deficiency of electrons. What happens when a neutral wire is connected between the charged plates? We

+ **14-4 CONDUCTORS**

Conductors and insulators

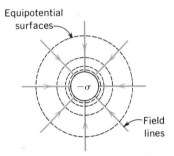

FIGURE 14-5 Spherical equipotential surfaces everywhere perpendicular to the radial electric field lines from a uniformly negatively charged spherical shell.

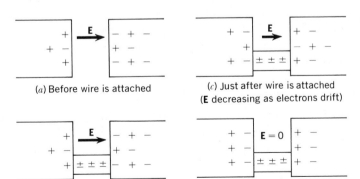

(a) Before wire is attached

(b) At instant wire is attached
(**E** also inside wire)

(c) Just after wire is attached
(**E** decreasing as electrons drift)

(d) Final condition

FIGURE 14-6 (a) Two parallel metal plates, the left with a deficiency and the right with an excess of negative charge. (b) The two plates at the instant an uncharged metal wire is connected between them. (c) A short time later the drift of charge within the plates and through the wire has reduced the imbalance of charge on the plates and the electric field between them. (d) Finally the imbalance of charge on the plates and the electric field between them are reduced to zero.

know that before the wire is connected, the opposite charges are mutually attracted. With the plates firmly anchored, however, the charges cannot meet. Although electrons are free to move within either plate, they cannot escape the forces at the surface which bind them in (unless the charges on either plate and the electric field between them are so large that a spark jumps across the gap). When a metal wire joins the plates, free electrons immediately start to drift in the wire toward the positive plate. After a very short time no net charge remains on either plate; we say that the plates have been discharged. This is not to imply that the excess electrons originally on the right plate have moved all the way to the other plate. It is rather a matter of many electrons making small shifts all along the wire. In materials other than metals both positively and negatively charged particles may be mobile. For example, in a conducting solution (an *electrolyte*) positive and negative ions move in opposite directions under the influence of an electric field.

Metallic conductors have special properties after the movement of charge in them has stopped. In this case a conductor is said to be in *electrostatic equilibrium*. A free electron will move inside a conductor in response to even a minute electric field. Thus, in the equilibrium state, the electric field in a perfect conductor is zero.

Electrostatic equilibrium: the electric field in a conductor is zero.

Moreover, no *net* charge can accumulate anywhere in the conductor; for if there were a net charge in some region, it would create a nonzero electric field. It follows that any net charge borne by a perfect conductor must reside on its surface. This charge must so distribute itself as to produce no electric field inside the conductor. For a sphere this means a uniform surface density of charge; as we know, the field of a uniform spherical shell of charge is zero at all interior points. For other shapes of conductor the surface charge is no longer distributed uniformly. Instead, it is most concentrated, and hence the electric field is strongest, at those points where the surface is most sharply convex, e.g., at a corner or any sharp point (see Fig. 14-7).

Net charge on a conductor resides on its surface.

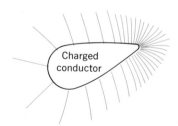

Just outside a charged conductor's surface the electric field is always perpendicular to the surface, as shown, for example, in Figs. 14-3, 14-5, and 14-7. Indeed, if it were not, at some point there would be a nonzero component of the electric field parallel to the surface, and free electrons would be set in motion, contradicting the original assumption of electrostatic equilibrium.

FIGURE 14-7   Electric field lines resulting from the nonuniform surface-charge distribution of a nonspherical metal object. The surface-charge density, and hence the density of electric field lines, is greatest where the surface is most sharply convex.

In summary, *for a perfect conductor in electrostatic equilibrium* (1) *all net charge is at the surface;* (2) *the electric field is perpendicular to the surface*

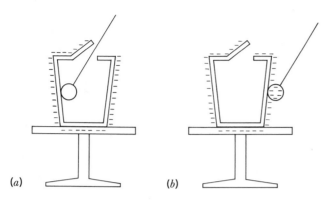

(*a*)   (*b*)

at each point and is exactly zero everywhere inside the conductor. According to the second property, no work is done on a charged particle by electric forces as the particle is transported from one point of the conductor to another. This leads to yet another characterization of a perfect conductor: (3) *the electric potential has the same value at each point inside and on the conductor.*

All these basic electrostatic properties of conductors are illustrated in the so-called ice-pail experiment of Michael Faraday (1791–1867). In this experiment, remarkable in its simplicity and in its far-reaching implications, Faraday placed a metal pail on an insulated stand, as shown in Fig. 14-8. The metal pail is then charged. To sample the charge at any one spot on the pail, one touches it with a small metal disk mounted on the end of an insulated handle; the amount and sign of the charge picked up by the sampling disk can then be measured with an electroscope of the sort described in Addendum 1. Faraday's findings were these: no charge is transferred to the small disk when it is touched to a point on the *inside* of the pail. This is the result even when the pail has originally been prepared by introducing charge to the inner surface. However, charge *is* detected if the disk is touched to the *outer* surface of the pail, and one finds larger amounts of charge near sharp edges.

The Faraday experiment illustrates *electrostatic shielding*. All electrostatic forces can be eliminated from a particular region of space simply by enclosing that region with a metal box. The interior is then cut off from the effects of all fixed charges outside the box. In fact, Faraday built a big cubical box covered with tinfoil and actually lived in it while performing a number of electric experiments. Even though, as he wrote, ". . . the outside of the cube was powerfully charged, and large sparks and brushes were darting off from

Electrostatic shielding

every corner of its outer surface . . . ," he was unable to detect any effect on himself or his sensitive measuring instruments. (One might well ponder whether there is an analogous gravitational shielding effect.)

Faraday's experiment also shows us how best to charge a large spherical conductor. The obvious method of touching smaller charged objects to the large conductor's exterior is not the most efficient. As the charge on the large sphere builds up, successive transfers become more and more difficult. Eventually a point is reached where the big sphere gives charge to the charged object touching it. The situation is rather like that of trying to fill a balloon with mouthfuls of air. Eventually, the pressure of the air in the balloon is so great that even though you have a mouthful of air, you simply cannot get it into the balloon. At still higher pressure, air would be transferred from the balloon to your mouth, not conversely.

Suppose, however, that the large sphere is provided with a small hole leading to an interior cavity. A small charged metal sphere can then be inserted into the cavity and touched to the interior wall. All excess charge from the small sphere is immediately transferred to the *outer* surface of the large metal sphere. The small sphere, now discharged, is pulled out, again touched to the battery terminal, reinserted into the interior of the large sphere, and again another "mouthful" of charge is transferred to the large sphere, first to its interior surface, but finally to its outer surface. Ideally, the process can be repeated indefinitely, for no matter how great the charge on the outer surface the interior of the cavity never acquires a charge. The only actual limitation to the total amount of charge that can be accumulated lies in the feeble conducting properties of the air or gas surrounding the big sphere. If the outside electric field becomes so large that the sphere arcs to ground or some other nearby object, the sphere loses its charge through the resulting sparks.

This charging process permits the creation of very strong electric fields, which can be used to accelerate charged atomic particles to great speeds. Such is the principle of the Van de Graaff generator described in Addendum 2 to this chapter.

EXAMPLE 14-5   Imagine that the two concentric spherical shells of charge in Example 13-9 (Fig. 13-17) are on the surfaces of thin *conducting* shells. What would happen to the distribution of this charge if the two conductors were connected by a radial conducting wire? Explain how your answer can also be obtained directly from Coulomb's law.

Because the shells and the connecting wire are conductors, effectively
constituting a single conductor, the net charge goes to the outermost
surface, i.e., the outer surface of the outer shell. We can arrive at
the same result differently. At the instant the wire is placed between,
and connected to, the two shells, the inward-directed electric field
will cause *electrons* in the wire to start to drift toward the outer shell. *Query* Why is an automobile a safe place in
This drift of electrons continues until the electric field is zero, i.e., a thunderstorm?
until all the net negative charge on the inner shell has moved to
the outer shell.

A negatively charged metal sphere is brought very close to one end EXAMPLE 14-6
of a long metal rod, initially uncharged (see Fig. 14-9a). The sphere
and rod are *not* in contact. (*a*) How is charge distributed over the
rod? (*b*) How would the charge distribution change if, with the
charged sphere still near one end of the rod, the other end is first
connected to the Earth (*grounded*) with a wire and then disconnected?
(*c*) What happens, finally, if the sphere is then moved away from
the rod?

a   When the negatively charged sphere is brought close to the metal SOLUTION
     rod, free electrons in the rod are repelled to the surface of the
     far end of the rod, leaving a net positive charge on the surface
     of the near end of the rod. (Of course, not all the net negative
     charge ends up exactly at the far end of the rod; the repulsive
     forces between the electrons, balanced by the forces from the
     charged sphere, keep the free electrons somewhat spread out; see
     Fig. 14-9a.)
b   If a wire connects the rod's far end to the ground, the electrons,
     still repelled by the negatively charged sphere, are pushed even

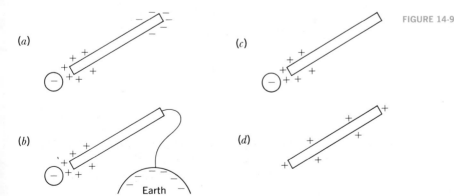

FIGURE 14-9

Charging by induction

farther away. The net negative charge is now spread thinly over the surface of the Earth (see Fig. 14-9b). If the wire is then disconnected, a net positive charge remains on the metal rod. We say that the rod has been charged by *induction* (see Fig. 14-9c). (Note that the charge induced on the initially neutral conductor has the opposite sign to that of the charged object.)

c Now if the sphere is moved away from the rod, some of the free electrons which kept the far end of the rod neutral (see Fig. 14-9d) will move back to the near end of the rod. As a result the positive charge at the near end is *partially* neutralized, and simultaneously a net positive charge appears at the far end. The rod now has a net positive charge distributed over its entire surface.

Query Is it necessary in charging the rod by induction to touch the ground wire to the far end? Could you touch it to the middle or to the near end?

SUMMARY

The change in electric *potential energy* for a system consisting of a particle interacting with two oppositely charged parallel plates is

[14-2] $\Delta E_p = F_e \, \Delta x$

where Δx is the component of the particle's displacement along a line parallel to the electric force, the positive x direction being opposite to the electric force (as in the gravitational analog, $\Delta E_p = mg \, \Delta h$). The *uniform* force $F_e = q(4\pi k_e \sigma)$ is perpendicular to the plates.

The change in the electric *potential*, defined as the electric potential-energy difference per unit charge, is measured in volts, and for the above parallel plates is given by

[14-6] $\Delta V = E \, \Delta x$

Electric field lines extend from positive charge to negative charge, the direction of decreasing electric potential.

For a perfect conductor in electrostatic equilibrium all charge is at the surface, the electric field is perpendicular to the surface at each point and is exactly zero everywhere inside the conductor, and the electric potential has the same value at each point inside and on the conductor.

ADDENDUM 1
THE ELECTROSCOPE

In the first paragraphs of Chapter 13 we showed that the electric force between suspended objects could be measured by balancing that force against weight and string tension, as shown in Fig. 14-10. A variation of this idea figures in the *electroscope*, a more accurate device which is often used in demonstrations. One simple type is shown in Fig. 14-11. It consists of a fixed metal rod AC slightly bent near its center B. A movable needle DE is pivoted at B. The needle, just slightly heavier in its lower half than in its upper, is easily displaced from its vertical equilibrium position. Once displaced, the needle quickly swings back to the vertical. Because of the bend, the upper and lower ends of the needle are always on opposite sides of the rod, as shown in Fig. 14-11. The rod passes through an insulating plug F, which in turn is supported by a metal ring G mounted on an insulating base H.

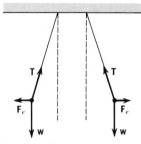

FIGURE 14-10   Detection of electric charge from the interaction between two light charged objects suspended from slender threads.

When the rod is touched at C by a charged object, a fraction of the charge is transferred down the rod and thereby to the needle of the electroscope. The needle immediately deflects because of the repulsion between the like charges on the surfaces of the needle and the fixed bar. This rotation is countered by the force of gravity, which is slightly greater on the lower half of the needle. The needle comes to equilibrium in the position at which these two competing effects just balance. Thus, a deflected needle indicates that the electroscope is charged; its angle of deflection is a measure of the amount of charge.

An electroscope cannot indicate the sign of the charge on it, at least not directly. The kind of charge, however, is easily determined by touching the top of the fixed rod with a small test object having a charge of known sign. If the needle's deflection increases, we know that the sign of the charge on the electroscope is the same as that on the test object.

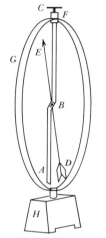

It is also possible to charge an electroscope without actually touching the electrified object to it. One simply charges the fixed rod by induction, in the manner described in Example 14-6.

FIGURE 14-11   Electroscope.

In 1931 R. J. Van de Graaff invented a machine for accumulating large amounts of charge on a conductor. The large electric fields and high electric potentials produced by the machine are especially useful in studies of nuclear structure. Van de Graaff incorporated into his machine the method of charging discussed in Section 14-4: charge is introduced through an interior cavity. The essential parts of the generator are shown in Fig. 14-12. The conductor to be charged is a large spherical metallic shell. Charge is carried on an endless fabric belt into the interior of the shell, through an opening at the bottom.

ADDENDUM 2

THE VAN DE GRAAFF GENERATOR

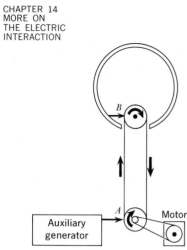

FIGURE 14-12 Basic principle of the Van de Graaff generator. Charge is transferred from a sharp needle A to the moving belt at the lower pulley. It is then transported by the belt to the interior of the spherical metallic shell, where it is transferred by a second sharp needle B to the shell. Because this charge moves to the outer surface of the shell, no strong electric field develops to oppose the transfer of additional charge.

The belt stretches over insulating pulleys, the lower one driven by a motor and the upper one, inside the spherical shell, driven by the belt itself. Charge is transferred to the belt as it passes over the lower pulley. This charge is brought to the interior of the spherical shell, where it runs past the very sharp end of wire needles which extend from the shell nearly to the belt. Strong electric fields are created between the charged belt and the induced charges at the tips of the needles (refer to Fig. 14-7). Ionization, i.e., the creation of charge carriers in the gas between the belt and the needles, occurs, and charge is transferred between belt and needles; but the net charge appearing on the shell immediately goes to the exterior spherical surface. As the belt keeps moving, charge is continually transferred to the shell. The upper limit to the amount of charge accumulated is determined mostly by the insulating properties of the gas around the shell and the legs supporting it. A final steady state is established when the charge being delivered by the belt is just matched by the charge leaking from the shell. Good commercial generators can attain electric potentials of the order of 10 million volts.

To charge the belt at the bottom, most systems use an auxiliary generator, which supplies charge at a constant electric potential. This source is connected to the belt by sharp needles, much like those at the upper end, but in simplified demonstration models of the generator, charge is developed by covering the lower pulley with a layer of wool and allowing the plastic belt to rub the wool as it rolls by. This is just a more sophisticated version of the very first electric experiment discussed in Chapter 13.

PROBLEMS

14-1   A large room 2.0 m high has positive charge spread uniformly over the ceiling and an equal amount of negative charge spread uniformly over the floor. A test object having a positive charge of 10^{-10} C experiences an electric force of 10^{-6} N when held anywhere in the room (away from the edges). What is the change in *kinetic energy* of the test object as it moves from the ceiling to the floor if its mass is (a) 9.1 gm, (b) 9.1×10^{-22} kgm, i.e. 1 billion times more massive than an electron?

14-2   What is the change in *total potential energy* as the test object of Problem 14-1 moves from ceiling to floor if its mass is (a) 9.1 gm and (b) 9.1×10^{-22} kgm?

14-3   What is the change in *electric potential energy* as the test object of Problem 14-1 moves from ceiling to floor if its mass is (a) 9.1 gm and (b) 9.1×10^{-22} kgm?

14-4 What is the change in *electric potential* as the test objects of Problem 14-1 move from ceiling to floor?

14-5 Two parallel plates are separated by a distance of 0.5 cm and charged to $\pm 3 \times 10^{-9}$ C/m². (*a*) What is the electric field **E** (magnitude and direction) between the plates? (*b*) What is the electric potential difference between the plates?

14-6 (*a*) An electron is accelerated through a potential difference of magnitude *1 V*. What is the increase in its kinetic energy? (*b*) Repeat this calculation using a proton.

14-7 An electron is projected straight upward with an initial kinetic energy of 0.2 eV through a small hole in the lower negatively charged plate of Problem 14-5. Will it reach the upper plate, and, if so, with what kinetic energy?

14-8 A proton of mass 1.67×10^{-27} kgm and charge 1.6×10^{-19} C is released from rest at the horizontal upper plate of two parallel plates. The negatively charged bottom plate is 0.02 m below the positively charged upper plate, and the electric potential difference between the two plates is 5.0 V. The plates have equal areas and equal magnitudes of charge. (*a*) What is the speed of the proton just before it reaches the lower plate? (*b*) What are the direction and magnitude of the uniform electric field between the two plates? (*c*) What is the magnitude of the charge per unit area on the plates?

14-9 A proton with a mass 1.67×10^{-27} kgm and charge $+1.60 \times 10^{-19}$ C moves with a horizontal velocity of magnitude 3.1×10^4 m/s as it enters the region between the two parallel horizontal plates shown in the figure. The plates are 0.02 m apart, and the proton enters at a point exactly midway between them. The bottom plate has an electric potential 5.0 V lower than the upper plate. Neglect gravity and assume the electric field to be uniform between the plates. (*a*) What is the vertical component of the proton's velocity when it hits one of the plates? (*b*) What is the total kinetic energy of the proton when it hits? Give your answer both in joules and in electron volts.

PROBLEM 14-9

14-10 The capacitance C of a two-conductor system is defined as the amount of charge per unit electric potential difference between the conductors that can be accumulated on one of the conductors by transfer from the second conductor. In symbols, $C = Q/\Delta V$, where Q is the magnitude of the amount of charge transferred from the one conductor to the other and ΔV is the electric potential difference between the two conductors. Show that the capacitance of two identical closely spaced parallel plates is equal to $A/4\pi k_e x$; that is, the capacity can be increased by increasing the area A of the plates *or* decreasing the plate separation x.

(a) C_1 C_2

PROBLEM 14-11, Part (a)

(b)

C_1

C_2

PROBLEM 14-11, Part (b)

14-11 (a) Show that if two sets of identical parallel plates are connected in parallel the capacitance C of the new two-terminal system is $C = C_1 + C_2$, where C_1 and C_2 are the individual values for the separate sets of plates. (b) Show that if the two sets of plates are connected in series, the capacitance of this new two-terminal system is given by $1/C = 1/C_1 + 1/C_2$. (See Problem 14-10.)

14-12 How much work is required to bring two protons to within 1.5×10^{-15} m of each other? (This is approximately the separation of two protons in a helium nucleus.) Give your answer in electron volts.

14-13 Two protons are released from rest 10^{-10} m apart. (a) What will their speeds be when they are 10^{-5} m apart? (b) What would your answer to this question be if one of the protons had been securely bound ("nailed") to a *massive* object (like a snowflake or the laboratory bench)?

14-14 Two protons have fixed positions 4.0×10^{-10} m apart. An electron moves along the direct line segment between the two protons, starting at an initial position 0.5×10^{-10} m from one proton and moving to a final position 1.0×10^{-10} m from the other proton. (a) What is the change in kinetic energy of the electron? (b) If the electron had an initial kinetic energy equal to this change, could it have reached its final position?

14-15 Two particles each having the same mass m_0, the same magnitude of charge q_0, but opposite signs of charge are released from rest when they are separated by an initial distance R. (a) What will be the speed of each particle when their separation distance is $R/2$? Assume they move freely. (Neglect gravity and give your answer in terms of m_0, q_0, k_e, and R.) (b) How will your answer to part (a) differ if the mass of one of the particles is $2m_0$ instead of m_0. *Hint:* What does conservation of momentum imply about the two speeds?

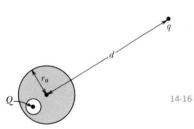

PROBLEM 14-17

14-16 Using the algebraic symbol q for charge reduce (a) [14-7] and [14-9] to one equation and (b) [14-8] and [14-10] to one equation.

14-17 A charge Q is placed inside a spherical cavity within an otherwise solid *metal* sphere of radius r_0. Another charge q is fixed a large distance $d \gg r_0$ away. (a) What is the net electric force on q? (b) What is the net electric force on Q?

14-18 An electroscope consists of a metal rod with a metal sphere at its upper end and two light gold leaves attached to its lower end. This system initially has a net charge of unknown sign. (a) A *negatively* charged rod is brought near the top sphere, and the leaves move closer together. Explain what is happening. What is the sign of the initial charge on the electroscope? (b) When the same negatively charged rod is used to touch the top sphere, the leaves first come together and then fly apart. Explain what is happening. What is the sign of the final charge on the electroscope?

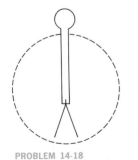

PROBLEM 14-18

Two concentric conducting spherical shells of radii r_1 and r_2 carry negative   14-19
charges $-q_1$ and $-Q_2$. (a) What is the electric field (magnitude and direction) at any point $r < r_1$, at $r_1 < r < r_2$, and at $r > r_2$? (b) Why is it
possible to transfer the charge from the inner shell to the outer shell by
connecting them with a wire but not from the outer shell to the inner?

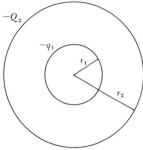

Consider a situation where a test object (charge q_t) is displaced along a   14-20
line making an angle θ with the electric field. Use the definition of electric
potential, the energy-conservation principle, and the work–kinetic-energy
relationship [8-13] to explain the following generalization of [14-6]:

PROBLEM 14-19

$$\Delta V = -E \cos \theta \, \Delta s$$

Note that for $\theta = 0$, **E** and **Δs** are in the same direction. Recall that for
[14-6] the positive x direction was opposite to the direction of the electric
force (field) in analogy to gravity.

THE MAGNETIC INTERACTION

The third of the basic classical interactions, the magnetic, was the last to be understood. Since ancient times it has been known that pieces of certain minerals could attract or repel each other and pick up bits of iron. The effects of these magnets seem to be concentrated near their ends, or *poles,* which should therefore be the carriers of the charge responsible for the magnetic force.

One of the prime discoveries in all of physics is that this is not so. Distinctive isolated magnetic charges have not been found in nature. Magnets, although they certainly illustrate the magnetic interaction, are really not required for magnetism. What *is* needed are *electrically charged particles in motion.* This means that magnetism is not really separate from electricity; the two interactions have a common origin.

> The magnetic interaction occurs between electrically charged particles in motion.

15-1 MAGNETS AND MOVING CHARGES

The crucial discovery that electric and magnetic effects are related was made by the Danish physicist, Hans Christian Oersted (1777–1851) in 1820. According to one story, Oersted wanted to show decisively in a lecture demonstration that electricity exerted no influence on a magnet. But, to his astonishment, he saw a compass needle *turn* when electric current was sent through a nearby wire. This fundamental experiment can be performed by anyone: a flash-light battery, a piece of wire, and a pocket compass are all that is needed (Fig. 15-1). The so-called *Oersted effect* was demonstrated even more strikingly by the American physicist Henry A. Rowland (1848–1901) in 1878. Without resorting to the relatively indirect procedure of having electrons move through the interior of a con-ducting wire, Rowland put an electric charge on an insulating disk (Fig. 15-2a) and set it spinning at a high rate (Fig. 15-2b). A freely pivoted magnet near the disk was deflected.

> The Oersted effect: electrically charged particles in motion can turn a magnet.

FIGURE 15-1 (*a*) A compass needle in the vicinity of a wire carrying *no* current points in a northerly direction. (*b*) When the wire is connected to a battery, creating a current in the wire, the compass needle is deflected.

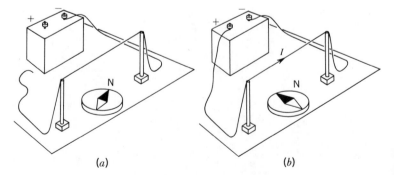

(*a*)   (*b*)

Thus, electrically charged particles *in motion* can turn a magnet. And it works the other way around: a magnet exerts a force on electrically charged particles in motion. To demonstrate this, carefully bring a magnet close to the face of the picture tube in a television receiver. Near the magnet the image becomes distorted. What happens is that the negatively charged electrons, coasting in at high speeds to collide with the face of the tube, are deflected from their orderly paths by a magnetic force arising from the magnet.

In these experiments, the magnet can be dispensed with and replaced by a second group of moving electric charges. Just a week after he had heard about Oersted's finding, André M. Ampère (1775–1836) showed that one conducting wire through which electrically charged particles were moving could be attracted or repelled by a second wire also carrying an electric current. Since each conductor was electrically neutral as a whole, the effect cannot be ascribed to the Coulomb, or electric, force previously discussed. Magnetism is thus basically associated with electric charges in *motion*.

Figure 15-3 sketches the fundamental discoveries in the evolution of our understanding of the magnetic interaction: in (a) is shown a fixed magnet causing a pivoted magnet (a compass needle) to move; in (b) moving electric charges in a fixed conductor causing

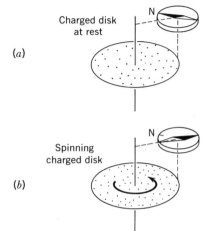

FIGURE 15-2 (*a*) A compass needle in the vicinity of a charged disk at *rest* points in a northerly direction. (*b*) When the disk is set spinning at a high rate, the compass needle is deflected.

FIGURE 15-3 Summary of magnetic effects.

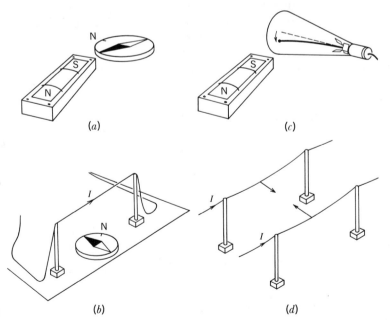

a pivoted magnet to turn; in (*c*) a fixed magnet deflecting a beam of electric charges, and in (*d*) we see the fundamental magnetic effect, the interaction between two (or two groups of) moving electric charges. We shall have a surprising entry to add to this sequence in Chapter 16.

15-2 THE MAGNETIC INTERACTION BETWEEN TWO MOVING POINT CHARGES

The interaction between two *moving* electric charges is unlike anything we have dealt with before. Consider two point charges q_1 and q_2 separated by the distance r. Charge q_1 has a velocity $\mathbf{v}_1$ and q_2 a velocity $\mathbf{v}_2$. We are interested in the total force on each charge. For the directions of the velocity vectors shown in Fig. 15-4*a* the forces on q_1 and q_2 are $\mathbf{F}_1$ and $\mathbf{F}_2$ shown in Fig. 15-4*b*. The magnitudes of $\mathbf{F}_1$ and $\mathbf{F}_2$ are not the same, and their directions are not opposite. Apparently Newton's third law does not describe this interaction.[1]

We describe part of the force on each particle by Coulomb's law, $\mathbf{F}_{e1}$ and $\mathbf{F}_{e2}$ in Fig. 15-4*c*. These two forces *are* equal in magnitude and opposite in direction.

The remaining part of the force on each particle, that part of the interaction which depends on the particle's motion, is the *magnetic force*, shown in Fig. 15-4*c* as $\mathbf{F}_{m1}$ and $\mathbf{F}_{m2}$. It is these *velocity-dependent* forces which are not of the same magnitude and do not act along the line connecting the two point charges. In short, the most general interaction between two electrically charged particles consists of two parts: (1) the electric part, which exists whether the particles are in motion or at rest, and (2) the velocity-dependent part, the magnetic force, which exists only when both charged particles are in motion.

[1] Moreover, if Newton's third law does not hold, it follows that linear-momentum conservation also does not apply to a pair of electric charges in general (see Section 4-2). Furthermore, the law of angular-momentum conservation would be invalid when applied to a pair of interacting electric charges (see Sec. 7-5). Suffice it to say here that the linear-momentum and angular-momentum conservation laws *do* apply to this electromagnetic interaction as soon as it is recognized that in addition to the linear momentum and angular momentum of the charges alone one must include the linear and angular momentum of the electric and magnetic fields themselves. More about this in Section 20-4.

The fundamental magnetic effect: the additional force between two moving electric charges, not accounted for by Coulomb's law, is the magnetic force.

Just for the record (and *not* to be memorized), the magnitude of the magnetic force exerted by charge q_1 moving with speed v_1 on charge q_2 moving with speed v_2 and separated from q_1 by the distance r is

$$F_{m2} = \frac{k_m(q_1 v_1 \sin\theta)(q_2 v_2 \sin\phi)}{r^2} \qquad [15\text{-}1]$$

FIGURE 15-4 (*a*) Two nearby positive electric charges moving in different directions (*b*) exert forces on each other which are not in general along the same line. (*c*) Part of each force, the electric component $\mathbf{F}_e$, is accounted for by Coulomb's law of interaction. The remaining part, the velocity-dependent component $\mathbf{F}_m$, can be accounted for by a magnetic interaction. The electric forces are along the same line, but this is not generally true for the magnetic forces.

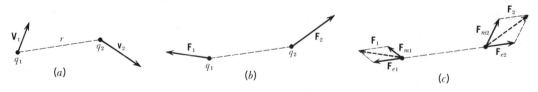

(*a*)   (*b*)   (*c*)

The angles θ and ϕ are identified in Fig. 15-5. The direction of the magnetic force follows a set of relatively complicated rules which we omit at this point. Soon we shall have a simpler way to describe this magnetic force.

The main point to be got from [15-1] is that the magnetic force, like the electric force, obeys an inverse-square law. The quantity k_m is a constant that gives a quantitative measure of the strength of the magnetic interaction between moving charged particles, just as the constant k_e in Coulomb's law measures the electric interaction. In the mks system of units the magnitude of k_m for two particles moving in *vacuum* is assigned the value

$$k_m = 10^{-7} \text{ N-s}^2/\text{C}^2 \qquad\qquad [15\text{-}2]$$

Normally, the magnetic force is very much weaker than the electric force, becoming comparable only when the charges move at speeds approaching that of light (3×10^8 m/s). To prevent the magnetic force from being totally masked, experiments usually have to nullify the electric force by using equal amounts of positive and negative charge and allowing only one kind to move. This was the case, for example, in Ampère's setup.

Magnetic force is inverse-square.

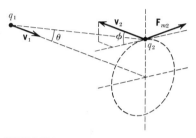

FIGURE 15-5 The magnetic force $\mathbf{F}_{m2}$ exerted on charged particle 2 by charged particle 1. The direction of this force is given relative to the directions of the particle velocities and the direction of the location of particle 2 with respect to particle 1. The magnitude of the force is given by [15-1].

15-3 ELECTRIC CURRENT, RESISTANCE, AND OHM'S LAW

Electric current defined

When charged particles move through empty space or through a conducting material, the amount of charge crossing a fixed surface per unit time is called the *electric current* through that surface. In symbols, if in time Δt a net charge Δq is transported through an area A, the current through A is given by

$$I = \frac{\Delta q}{\Delta t} \qquad\qquad [15\text{-}3]$$

Since charges of either sign can pass through A from either side, we need in addition to [15-3] an agreement about what *direction* to associate with an electric current. The convention is this: the direction of an electric current is that in which *positive* charges move (see Fig. 15-6). Thus, in a straight conductor, positively charged particles moving to the right constitute an electric current to the right, whereas negatively charged particles also moving to the right (the equivalent of positive particles to the left) constitute an electric current to the left. Another, equivalent way of fixing the current direction is that of the electric field which drives the charged particles. Using either rule, we readily see that in an ordinary wire, in which the charge

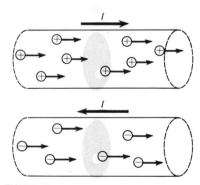

FIGURE 15-6 In a straight conductor (a) positive charges moving to the right constitute a current to the right, and (b) negative charges moving to the right constitute a current to the left. The magnitude of the current is given by [15-3].

carriers are electrons, the conventional current is always opposite to the direction in which the electrons actually drift.

The units of electric current are, from the definition, those of charge divided by those of time. In the mks system the combination coulomb per second is called the *ampere* (A). Whereas 1 C is an uncommonly large charge, 1 A of current is quite typical (a household appliance may well take 1 A). Laboratory experiments, however, often involve currents in the microampere range (μA, 10^{-6} A).

Current unit, the ampere (A)

To investigate the magnetic effects of an electric current we want to relate the current to the properties of the individual charge carriers. For a cylindrical conductor this is easy. Assume that each carrier has a charge q and drifts with constant speed v_1 parallel to the axis of the cylinder.[1] Let the cross-sectional area be A_1 and the density of charge carriers (the number of charged particles per unit volume) be n_1. We concentrate on the charge carriers in the cylindrical element of length $v_1 \Delta t$, as shown in Fig. 15-7. After a time interval Δt has elapsed, all these carriers will have moved through one end of the little cylinder. Now the total charge within the element is its volume $(v_1 \Delta t)A_1$ times the number of carriers per unit volume n_1 times the charge per carrier q. The definition of electric current then gives

[1]Of course the charge carriers (unbound electrons) are in chaotic thermal motion, like the molecules of an ideal gas. Upon this random motion is superposed an orderly drift, under the influence of the driving electric field inside the conductor. Since only the drift component produces a net transport of charge, we are justified in representing the carriers' motion by the average drift speed v_1.

$$[15\text{-}4] \quad I = q n_1 v_1 \Delta t \frac{A_1}{\Delta t} = n_1 q v_1 A_1$$

EXAMPLE 15-1

About how fast do electrons drift through an ordinary household wire when a typical electrical field is applied?

SOLUTION

Suppose that a current of 1.0 A exists in a copper conducting wire with a diameter of 1.0 mm. Then the cross-sectional area is $A_1 = \pi(0.50 \text{ mm})^2 = 7.9 \times 10^{-7}$ m^2. The charge of each electron is $q = e = 1.6 \times 10^{-19}$ C. In copper there is on the average approximately one conduction electron for each copper atom. The density of copper atoms is 8.4×10^{28} atoms/m^3, hence, $n_1 = 8.4 \times 10^{28}$ electrons/m^3. Substituting these values into [15-4] and solving for the drift speed v_1, we find that $v_1 = 9.4 \times 10^{-5}$ m/s = 0.094 mm/s. Surprisingly, perhaps, the electrons drift through the conductor at a very low speed, less than $\frac{1}{10}$ mm/s. This should not be confused with the very high speed (close to that of light) with which the driving electric field (and hence the current) is set up after the switch is closed.

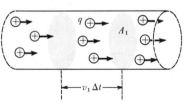

FIGURE 15-7 If all the charges in a beam of cross-sectional area A_1 are moving with speed v_1, the total charge in an element of the beam of length $v_1 \Delta t$ will be carried past a fixed cross section through the beam (or conductor) in a time Δt.

It is also possible to relate the current directly to the driving field. For such a field to exist, an electric potential difference ΔV must be maintained across the conductor between the points where the electric charges enter and leave (see Fig. 15-8). It was discovered by Georg Ohm (1787–1854) that for most common solid conducting materials the current I is directly proportional to the potential difference ΔV

$$\Delta V = RI \qquad [15\text{-}5]$$

where R is a constant, characteristic of the conductor, called the conductor's *resistance*. Relation [15-5] is known as *Ohm's law*. Although quite general, it does not hold, for example, for semiconductors or ionized gases.

The resistance of a conductor carries the units of volts per ampere; a special name, the ohm (abbreviated Ω), is given to this combination of units.

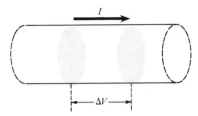

FIGURE 15-8 The electric potential difference ΔV between two points along a conductor carrying a current I.

Electric resistance defined

There are two reasons for expecting that the situation to be investigated in this section will provide basic information about the magnetic force: (1) Our conductor as a whole is electrically neutral; for every negatively charged conduction electron moving through the wire there is a corresponding positive ion nearby which remains fixed in the solid. There can, then, be no resultant *electric* force between the conductor and a charged particle outside it. *Magnetic* effects alone, arising from the moving charges in the conductor, will influence the motion of the outside particle. (2) For this simple geometrical arrangement it is possible to treat the moving charges in the fixed conducting wire as though they traveled in straight lines at a constant drift speed.

15-4 MAGNETIC INTERACTION BETWEEN A MOVING POINT CHARGE AND A LONG STRAIGHT CURRENT-CARRYING WIRE

Since as a detector of magnetic force a single moving charge is hard to use, we shall instead use a whole beam of them, say the electron beam in a small TV picture tube. First adjust the beam so that in the absence of any external magnetic field a single bright spot is produced at the center of the tube face. This shows that the electrons are moving straight down the axis of the tube. The presence of an applied force can be detected by the deflection of the bright spot from its central position.

When such a detector is placed near a long straight current-carrying conductor, the following observations are made:

Only one line along which electron beam can be fired and experience no magnetic force

1 Through any given location near the conductor there is one special line along which we can orient the electron beam and find no force on it at the given point. This line, shown as the z' axis in Fig. 15-9a, is always tangential to a circle drawn through the location of the test object (here, a short segment of the beam) and concentric to the straight conductor. Indeed, the electrons can move in either direction along this line, as shown by the velocity vectors **v** in Fig. 15-9a and b respectively, and still experience zero magnetic force.

2 When electrons move in the plane both perpendicular to the above direction of zero force and containing the wire, they experience (for a given speed v and a given current I) a magnetic force of maximum magnitude F_{max}. For example, if the electrons move, as shown in Fig. 15-9c, in the same direction as the current in the wire, they experience this maximum force. In every instance, this maximum magnetic force is directed along a line perpendicular to *both* the line of zero force and the electron velocity vector. (Check this for the special case shown in Fig. 15-9c and the more general case shown in Fig. 15-9d.)

Force is maximum when beam is fired at right angles to this line.

3 This last rule, of course, does not describe completely the direction of F_{max} since there are two possibilities along the specified line. This ambiguity will be resolved in the next section, when we shall be able to use the simplifying concept of a magnetic field. Suffice it for now to say that the direction of the magnetic force is reversed when any one of the following changes is made: the current direction in the straight conductor is reversed (compare Fig. 15-9c and e), the test object's velocity direction is reversed (compare Fig. 15-9c and f), or the sign of the test object's charge is reversed (compare Fig. 15-9c and g).

4 Thus far we have considered the magnetic force only in those cases where it has its maximum magnitude, i.e., when the electron moves perpendicular to the line of zero force. Happily, the simple rule (paragraph 2) still holds when the velocity vector of the moving test object makes an arbitrary angle ϕ with the line of zero force (Fig. 15-9h). That is to say, the line of the magnetic force is perpendicular to both the velocity vector of the test object and the line of zero force. But the magnitude F_m of the magnetic force now depends on the value of the angle ϕ, according to

[15-6] $$F_m = F_{max} \sin \phi$$

5 Finally, if the amount of charge q_2, or the velocity v_2 of the test object, or the current I in the conductor is changed, it is found that the magnitude of the magnetic force changes in direct pro-

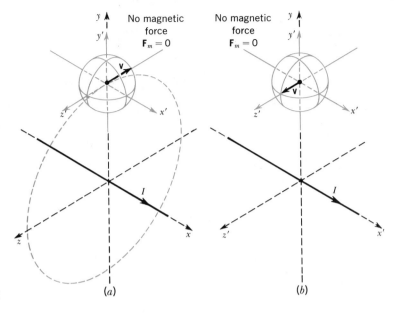

FIGURE 15-9, Part (*a*)   A point charge in motion near a long straight current-carrying wire experiences no magnetic force if the point charge moves along a line that is *tangent* to a circle drawn through the location of the charge and concentric to the wire. Part (*b*)   If the velocity of the charge in (*a*) is reversed in direction, the magnetic force remains zero.

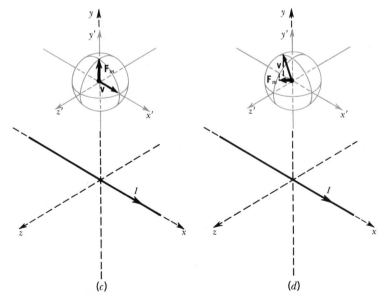

FIGURE 15-9, Part (*c*)   When the charge moves parallel to the wire, the magnetic force is in the radial direction. Part (*d*)   When the charge moves in the plane of the wire, i.e., perpendicular to the tangent line of zero force shown in (*a*), it experiences a magnetic force in a direction perpendicular to both its velocity and the line of zero force. Moreover, for a given speed, radial distance from the wire, and current, this is the magnetic force of *maximum magnitude.*

FIGURE 15-9, Part (e)   If the current direction in (c) is reversed, the direction of the magnetic force is also reversed. Part (f) If the velocity direction of the charge in (c) is reversed, the direction of the magnetic force is also reversed.

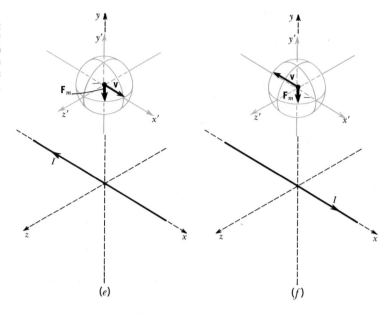

(e)   (f)

FIGURE 15-9, Part (g)   If the sign of the charge in (c) is reversed, the direction of the magnetic force is also reversed. Part (h) In the most general case (the velocity no longer restricted to the plane of the wire perpendicular to the tangent line of zero force and the force no longer necessarily of maximum magnitude) the magnetic force $\mathbf{F}_m$ is still perpendicular to both the velocity and the tangent line of zero force (z' axis). The magnitude of this force is now given by [15-6]. Here ϕ is the angle between the velocity and the tangent line of zero force.

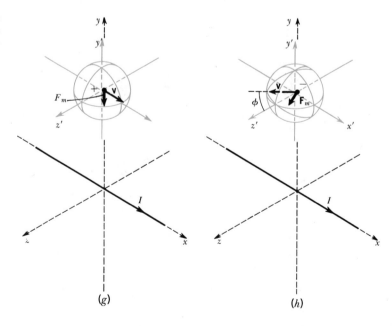

(g)   (h)

portion. On the other hand, it is found that the force is inversely proportional to the radial distance d of the test object from the straight conductor. Using the same magnetic constant k_m given in [15-1], we can rewrite [15-6] as

General relation for the magnetic force on a particle near a long straight wire conductor

$$F_m = k_m \frac{2I}{d}(q_2 v_2 \sin \phi) \qquad [15\text{-}7]$$

This equation, together with the direction rule, gives us the essential information but in an indigestible form. For a simpler and more powerful description of magnetic interactions we shall have to switch from the force viewpoint to the field viewpoint.

15·5 THE MAGNETIC FIELD

The idea of a magnetic field is like that of an electric field. We say that a moving charge q_1 creates a magnetic field in space surrounding it, and a second moving charge q_2 immersed in this magnetic field may be acted upon by a magnetic force. By regarding the magnetic interaction as taking place via the intermediary of the magnetic field, we can greatly simplify our description of the magnetic interaction.

In the electric case we can, at each point, directly represent the electric field by the electric force on a suitable test charge. For the magnetic field the procedure is not quite so simple. When exploring the magnetic field, we must use a charged particle moving with velocity **v** as the test object because the magnetic interaction depends on the velocities of the interacting particles. We think of the unit magnetic test object as being a unit positive charge moving at a unit speed. For the present let us restrict ourselves to the field of a long straight current-carrying conductor, as shown in Fig. 15-9.

1   The magnitude B of the *magnetic field* at any point is the maximum magnitude F_{max} of the magnetic force exerted on a unit test object at that point. (As we have noted, the magnitude of the magnetic force at any particular location differs according to the direction of the velocity; to have a maximum magnetic force the velocity vector of the test object must lie somewhere within the plane perpendicular to the unique line of zero force.) If the charge and speed of the test object are q_2 and v_2, then

Magnetic-field magnitude B defined

$$B = \frac{F_{max}}{q_2 v_2} \qquad [15\text{-}8]$$

2   *At any point in space* **B** *is oriented along the unique line of zero force.*[1] This line, as we know, lies in a plane perpendicular to the wire and is tangent to a circle centered on the wire.

[1] The line of the magnetic field, as defined here, is also the line along which a freely pivoted compass will align itself. More about this in Section 15-12.

3   To make the direction of **B** perfectly definite we have still to choose between the two senses along the tangent line. The conventional choice is shown in Fig. 15-10. A mnemonic for this convention is: *let the right thumb point in the direction of I; then the curled fingers of the right hand give the sense of the magnetic field* (see Fig. 15-11). This is our *right-hand rule for magnetic fields*.

The magnetic field outside a long straight wire carrying a steady current I can now be completely described. Just as for the electric field, there is a commonly accepted way of graphically displaying both the direction and the magnitude of the magnetic field **B**, that is, through the use of magnetic field lines. The *magnetic field lines*, i.e., the curves whose tangent at each point gives the direction of the vector **B** at that point, are circles centered on the wire and lying in planes perpendicular to it (see Fig. 15-12). At each point of a magnetic field line of radius d, the magnitude of **B** is (from [15-8] and [15-7] with $\phi = 90°$)

$$B = \frac{F_{\max}}{q_2 v_2} = k_m \frac{2I}{d}$$
[15-9]

and the direction of **B** is fixed by the right-hand rule (see Fig. 15-11).

Where the lines are crowded closely together, B is large; where the lines are sparse, B is small. As shown in Fig. 15-13, far from a straight conductor the lines are sparse relative to their spacing near the conductor. This simply indicates that the field grows weaker with distance away from the conductor. Also note that over relatively small regions of space far from the conductor neither the density nor the direction of the lines changes very much. We say that in such regions the field is very nearly uniform; its lines are very nearly straight, parallel, and equally spaced.

Note an important difference between electric and magnetic field lines. Sometimes electric field lines are called electric *lines of force* because the electric force is also always *tangential* to the electric field line. Not so for the magnetic field lines. The magnetic force is always *perpendicular* to the magnetic field. Thus we can never call a magnetic field line a magnetic line of force.

The units of the magnetic field are determined by [15-8] above as N/(C)-(m/s) = N/A-m. For brevity this combination of units is called the tesla (T).[1] As magnetic fields go, a field of 1 T is very large. For example, an ordinary bar magnet may produce a field of only 10^{-2} T. It is, therefore, convenient to use another smaller unit for B, called the gauss (G), where

Right-hand rule for magnetic *fields*

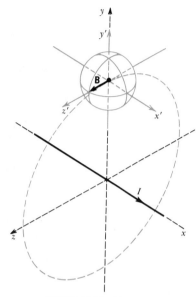

FIGURE 15-10   The direction of the magnetic field **B** for the position of the charge near the current-carrying wire in Fig. 15-9. The direction is along the tangent line of zero force in the sense shown.

FIGURE 15-11   The right-hand rule for magnetic fields, a mnemonic for the sense chosen in Fig. 15-10 for the direction of the magnetic field.

[1] It will be seen in Chapter 16 that the product $B \times$ area is as physically significant as B itself. This leads to an alternate unit for B, the weber per square meter (1 T = 1 Wb/m²).

Magnetic field unit, the tesla (T)

1 T = 10,000 G

At the surface of the Earth, the Earth's magnetic field is of the order of 1 G.

Our defining rules for **B,** though formulated with a special case in mind, actually hold for any distribution of moving charges. An important property of our special field extends to the general case. *The magnetic field lines are always closed loops* (although not necessarily circles). This is in contrast to electric (static) field lines, which always begin and end on charges. It is also true that the field **B** at any point is always directly proportional to the strength of the source current. For example, a single charge q moving at speed v_1 produces a magnetic field **B** whose magnitude is (from [15-8] and [15-1] with $\phi = 90°$)

$$B = \frac{F_{\text{max}}}{q_2 v_2} = \frac{k_m q_1 v_1 \sin \theta}{r^2} \qquad [15\text{-}10]$$

The magnetic field lines for a single moving charge are shown in Fig. 15-14. Again they are circles, now concentric about the charge's velocity. The magnitude of the magnetic field decreases with increasing distance from the charge in all directions, including the forward and backward directions.

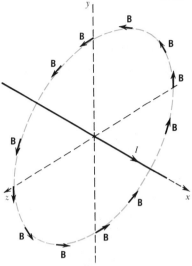

FIGURE 15-12   The directions of the magnetic field **B** in a plane perpendicular to a long straight current-carrying wire. All the points illustrated are equidistant from the wire.

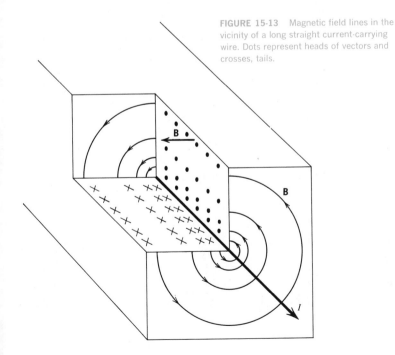

FIGURE 15-13   Magnetic field lines in the vicinity of a long straight current-carrying wire. Dots represent heads of vectors and crosses, tails.

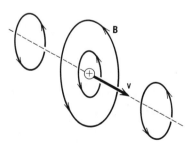

FIGURE 15-14   Magnetic field lines in the vicinity of a moving charged particle.

15-6 THE MAGNETIC FORCE

What is, in general, the direction and magnitude of the magnetic force $\mathbf{F}_m$ on a charge q_2 moving with velocity $\mathbf{v}_2$ in a magnetic field **B**? We know, from the definition of **B**, that the force is zero when $\mathbf{v}_2$ is parallel (or antiparallel) to B and that F_m is maximum for any direction of $\mathbf{v}_2$ perpendicular to **B**. In general, [15-6] holds, and $F_m = F_{\max} \sin \phi$, where ϕ is the angle between $\mathbf{v}_2$ and **B.** But we also know from the definition, [15-8], of **B** that the maximum force ($\phi = 90°$) is $F_{\max} = q_2 v_2 B$. Thus when a charged particle moves through a magnetic field, the *magnitude* of the magnetic force on it is

Magnetic force on a moving charge

$$F_m = q_2 v_2 B \sin \phi \qquad [15\text{-}11]$$

Since the component of $\mathbf{v}_2$ at right angles to **B** is $v_\perp = v_2 \sin \phi$, [15-11] can also be written as

$$F_m = q v_\perp B \qquad [15\text{-}12]$$

Here we have also dropped the subscript 2, as might well be done in [15-11], for now we implicitly understand that q and v refer to the particle on which the force F_m acts (in contrast to the particles which are the source of B).

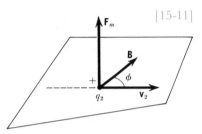

FIGURE 15-15 Direction of the magnetic force $\mathbf{F}_m$ on a particle of positive charge q_2 moving with velocity $\mathbf{v}_2$ in a magnetic field **B**. The force $\mathbf{F}_m$ is perpendicular to the plane of $\mathbf{v}_2$ and **B**, and is directed as shown.

The observed direction of $\mathbf{F}_m$ is shown in Fig. 15-15 (positive q_2). The magnetic force is always at right angles to the plane containing the vectors **v** and **B.** To remember its direction we can use the following *right-hand rule for magnetic forces:* let the *right-hand* thumb point in the direction **v** and the index finger in the direction of **B** (the angle between the thumb and the index finger is less than 180°); then the middle finger of the right hand, when perpendicular to the first two fingers, points in the direction of $\mathbf{F}_m$ (see Fig. 15-16). This holds for a *positive* charge; the force direction on a moving negative charge is simply the opposite direction.

Right-hand rule for magnetic *forces*

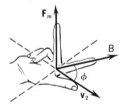

FIGURE 15-16 The right-hand rule for magnetic forces, a mnemonic for the direction of the magnetic force in Fig. 15-15.

The magnetic force is peculiar in several respects. It exists only when both of two charged particles are in motion. Moreover, since the magnetic force always acts at right angles to a charged particle's velocity, it can do no work. Thus, a magnetic field can deflect a particle but cannot change its kinetic energy. With such a "workless" interaction force, there is no way to associate a simple potential energy with the magnetic force. The most remarkable aspect of all is treated in the Addendum to this chapter: the magnetic force can be made to vanish or come into existence by a simple change of reference frame!

A magnetic force does not do work.

EXAMPLE 15-2

A *flexible* circular loop of wire, capable of stretching or shrinking, is placed perpendicular to a magnetic field. As shown in Fig. 15-17, the magnetic field is *into* the paper. (We use ✕'s to represent such a field, since the ✕'s remind us of the feathered tails of arrow vectors going into the paper. To show a field emerging out of the paper we use dots, which remind us of the points of arrows coming out of the paper.) Now a current in the *counterclockwise* sense is turned on in the loop; i.e., we imagine positive charges to circulate counterclockwise around the loop. Will the loop shrink or expand?

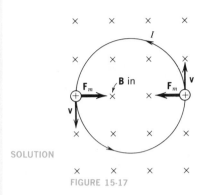

FIGURE 15-17

SOLUTION

If we concentrate for the moment on a positive charge moving upward on the right-hand side of the loop and apply the rule for finding the magnetic-force direction, we find that the magnetic force on such a charge is toward the left. The moving charge must remain in the conducting wire, so that the force is transmitted to the wire itself. By the same token, the force on a charge moving downward on the left side of the loop is toward the right. In general, the magnetic force on each segment of the loop is toward the center of the loop. Therefore, the loop shrinks. Of course, if the current direction were reversed, the loop would expand.

Query We know that in metal wire loosely bound negatively charged electrons move and the positive ions are stationary. Starting with this more basic fact, show that the answer to Example 15-2 is still correct.

15-7 CHARGED PARTICLES IN A UNIFORM MAGNETIC FIELD

For a uniform, or constant, magnetic field, the field lines are straight, parallel, and uniformly spaced. Any one location in such a field is like any other. To find how a charged particle injected into such a field moves, let us examine two extreme cases: (1) If the particle is injected with velocity **v** parallel or antiparallel to the lines of **B,** then, by the definition of **B,** it is subject to no magnetic force. Therefore, the particle coasts in a straight line with constant speed, as shown in Fig. 15-18a. (2) If the particle is injected with velocity **v** at right angles to the magnetic field lines, it experiences a magnetic force of magnitude qvB. Since the magnetic force can do no work, the particle's kinetic energy does not change and qvB remains constant. We know that a force of fixed magnitude perpendicular to the velocity produces motion in a circle at constant speed. In this case the circle will be in a plane perpendicular to the lines of **B** (see Fig. 15-18b). Its radius can be found from Newton's second law

Particles injected at right angles to a uniform magnetic field move in circles.

$$\mathbf{F} = m\mathbf{a}$$

$$qvB = \frac{mv^2}{r}$$

$$r = \frac{mv}{qB}$$

[15-13]

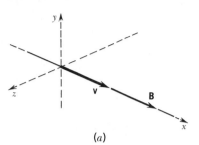

(a)

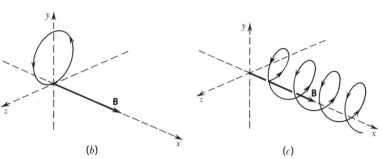

(b)   (c)

FIGURE 15-18   In a region where the uniform magnetic field **B** is everywhere in the direction of the positive x axis, (a) a charged particle initially shot in a direction parallel to **B** will continue to move in that direction at constant speed; (b) a charged particle initially shot in a direction perpendicular to **B** will move in a circle at constant speed; and (c) a charged particle initially shot in some other direction making an angle with **B** will trace out a helical path. (d) Bubble-chamber photograph of electron path. The electron enters the chamber of liquid hydrogen from the left. Pressure on the hydrogen has just been released, and the liquid forms bubbles readily about the ionized atoms produced by the rapidly moving electrons. The resulting track of bubbles marks the path of the electrons. A magnetic field applied perpendicular to the chamber makes the electrons move in circular paths, but because the electrons slow down as they ionize the atoms of the liquid, the radius of curvature of the circular path decreases and the electrons spiral in toward the center. (*University of California, Lawrence Radiation Laboratory, Berkeley.*)

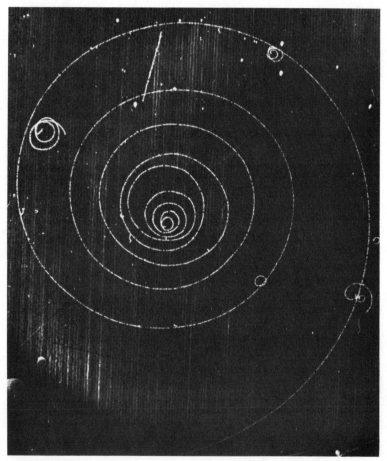

(d)

If for some reason the particle gradually loses energy and slows down, the radius of the path decreases and it spirals into the center (see Fig. 15-18d).

Suppose now the particle is injected neither along the field lines nor at right angles to them. What is its path? At any instant the velocity vector can be replaced by its components, one along **B** and the other at right angles to **B.** Insofar as the motion along **B** is concerned, the particle coasts at constant speed in a straight line. The perpendicular velocity component results in circular motion at constant speed. Thus, the overall motion, which is simply the superposition of these two motions, is along a helix, the curve one gets by wrapping a string around a right-circular cylinder with equal spacing between adjacent turns (see Fig. 15-18c).

General path of particle in uniform magnetic field: a helix

Consider a collection of identical charged particles, such as electrons, injected into a uniform magnetic field. The particles differ in their initial directions and speeds. Apart from those which happen to start off along magnetic field lines, all particles will, on the basis of the arguments given above, orbit the field lines in helical paths. The paths will differ in radius according to the speeds of the particles, but it is a remarkable fact that *all* particles will take exactly the *same time* to complete one loop around the field lines. That is, all particles will encircle the magnetic field lines at the same frequency.

This is easy to prove. Let f represent the frequency of a particle's circular motion. During one orbit, a particle travels at transverse speed $v_\perp$ around a circle of circumference $2\pi r$. The time taken is thus $2\pi r/v_\perp$, and this is the period of the motion. Since the frequency, or number of orbits per unit time, is just the reciprocal of the period,

$$f = \frac{1}{T} = \frac{v_\perp}{2\pi r}$$

[15-14]

Solving [15-13] for $v_\perp/r$ and substituting this result into [15-14], we have

The cyclotron frequency

$$f = \frac{1}{2\pi}\frac{qB}{m}$$

[15-15]

As [15-15] shows, this frequency (often termed the *cyclotron frequency* for reasons to be seen in Example 15-6) is proportional to the magnitude of **B** and to the particle's charge-to-mass ratio q/m, but f is independent of the particle's speed or the radius of the helix!

EXAMPLE 15-3   As a proton moves through a chamber containing liquid hydrogen near its boiling point, it leaves a trail of charged particles in its wake. On these ions as centers small bubbles form in the liquid, thus rendering the path of the proton visible. The device, called a *bubble chamber,* is much used in high-energy, or particle, physics.

What is the momentum of a proton found to be moving in a circular path of 5.0 cm radius in a uniform magnetic field of 0.10 T (1,000 G)?

SOLUTION   From [15-13] we have

$$mv = qrB = (1.6 \times 10^{-19})(5.0 \times 10^{-2})(0.10)$$
$$mv = 8.0 \times 10^{-22} \text{ kgm-m/s}$$

EXAMPLE 15-4   The earth's magnetic field was first mapped, at least roughly, by William Gilbert in 1600. It is as shown in Fig. 15-19, with the magnetic field lines emerging near the geographic South Pole and entering near geographic North. In magnitude the field is slightly less than 1 G near the surface and still smaller at distances farther out. Terrestrial magnetism is believed to have its origin in the circulation of electrically charged particles in the Earth's metallic inner core.

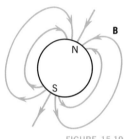

FIGURE 15-19

Cosmic rays

The magnetic field of the Earth influences the paths of charged particles from outer space, mostly protons, which rain upon the Earth continuously from all directions and are called *cosmic radiation.* Particles approaching the Earth along the polar axis move along magnetic field lines and are consequently undeflected (see Fig. 15-20a). On the other hand, particles approaching the Earth at the

FIGURE 15-20

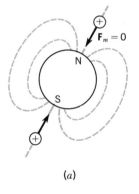

(a)

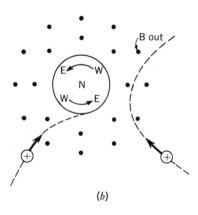

(b)

equator move at right angles to magnetic field lines and may be so strongly deflected that they miss hitting the Earth's atmosphere entirely. The intensity of the cosmic radiation is then greater near the poles than near the equator. Moreover, those positively charged particles which approach the Earth at the equator are deflected by a magnetic force directing them *toward* the east, and experiment shows that the intensity of radiation is greater *from* the west than *from* the east (see Fig. 15-20*b*).

The east-west effect shows that the primary cosmic-ray particles are positively charged.

Actually, the arguments go the other way around. *Because* the cosmic-ray intensity is found to be greatest at the poles, and because the intensity at the equator is greatest from the west, it is inferred that the primary particles of the cosmic radiation are, in fact, positively charged.

In 1958 the Van Allen belts, consisting of high-intensity radiation trapped in the Earth's magnetic field, were discovered in experiments with space probes supervised by J. A. Van Allen. There are actually two such belts, one about 2,500 miles from Earth's surface at the equator and a second one at about 11,000 miles. Figure 15-21 shows the general features of one such belt. Charged particles, mostly electrons, spiral around the Earth's magnetic field lines in paths which although not strictly helical are nearly so, because the magnitude and direction of the Earth's magnetic field change only gradually from one location to another nearby. When a charged particle approaches geographic north or south, where the field lines converge, it may be reflected to travel back toward the center of the belt and thence to the other extremity. The electrons are trapped in what has been called a *magnetic bottle*. The container holding the particles is merely the magnetic field. Such magnetic bottles, which, of course, will not melt at high temperatures, are of considerable practical interest to workers developing the useful release of energy in nuclear-fusion processes (Section 27-7).

Van Allen belts

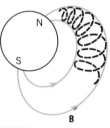

FIGURE 15-21

A magnetic bottle

A particle with charge q is to be sent through a combined uniform electric field **E** and a uniform magnetic field **B** *without* being deflected at all. How are the fields to be arranged, and what are their relative magnitudes?

EXAMPLE 15-5

Since an electric field acts on a charged particle, whatever the state of the particle's motion, the only way of eliminating the effect of the electric field is to balance the electric force against the magnetic force. We know that the magnetic force is always at right angles to

SOLUTION

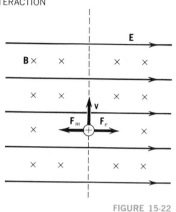

FIGURE 15-22

Crossed electric and magnetic fields as a velocity selector.

the particle's velocity. Therefore, the electric force, and hence the electric field, must also be at right angles to **v.** For example, with positively charged particles moving upward, as in Fig. 15-22, and with an electric field to the right, the magnetic field **B** must be directed into the paper, as shown. The magnetic force $F_m = qvB$ is to the left. The electric and magnetic forces just balance each other when

$$F_e = F_m$$
$$qE = qvB$$
$$v = \frac{E}{B}$$

[15-16]

There is a single speed, given by E/B, at which *any* charged particle (quite apart from its mass or the magnitude or sign of its charge) can pass through crossed electric and magnetic fields without being deflected. Such an arrangement, often called a *velocity selector,* can then be used to measure the speeds of particles or to select from particles having a variety of speeds those particular particles moving at the speed which satisfies [15-16].

EXAMPLE 15-6

The cyclotron, a high-energy particle accelerator

Side view

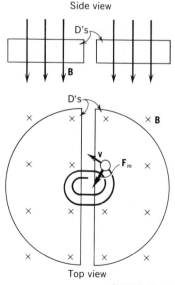

Top view

FIGURE 15-23

The cyclotron is a device for accelerating charged particles to relatively high kinetic energies. Unlike the Van de Graaff accelerator, in which the particles are accelerated along a straight line by *one* large electric potential difference, the cyclotron gives the particles *multiple* accelerations. It does so by employing a magnetic field through which the particles move at right angles to the field lines. The essential point is that a particle will travel in circular arcs around field lines at a frequency, the so-called *cyclotron frequency,* which is *independent* of the particle's speed or the radius of its path. As [15-15] shows, $f = qB/2\pi m$, where q and m are the particle's charge and mass and B is the magnitude of the magnetic field.

The basic parts of a cyclotron are shown in Fig. 15-23. Charged particles, such as protons, are released near the center of the magnetic field and start to move in a circular arc. Actually, the particles move inside the D-shaped hollow conductors between which an electric potential difference can be applied. While inside either of the two D's, a particle is completely shielded electrically and travels in a circular arc at constant speed, but when the particle enters the gap between the two D's, it becomes subjected to the accelerating electric field associated with the potential difference. It enters the interior of the next D at a higher speed, travels a half-circle of larger radius, and again emerges into the gap between the D's. An adjustment to

329

SECTION 15-8
THE MAGNETIC
FORCE ON
CURRENT-CARRYING
CONDUCTORS

be described below ensures that each time the particle crosses the space between the D's the electric field accelerates it. Thus it spirals outward and finally strikes a target near the exterior of the cyclotron with a very high energy.

It is required that the electric potential difference appearing across the D's alternate with time, so that the particles are always speeded up, rather than alternately speeded up and slowed down. This is, in fact, easy to achieve since the time taken by an orbiting charged particle to make one-half turn is always the same, independent of its energy or radius. That is, the cyclotron will operate properly if the frequency of an alternating electric potential difference applied to the D's is just the cyclotron frequency $f = qB/2\pi m$.

The cyclotron was invented by E. O. Lawrence and M. S. Livingston in 1932. Such a machine can, for example, impart a kinetic energy of 25 million electron volts (MeV) to protons by accelerating them 500 times across a potential difference of 50,000 V.

Before considering the moving particles confined in a *wire*, let us consider a *beam* of free particles. For simplicity we choose a beam having a rectangular cross section of area A_2 (see Fig. 15-24). The *total charge* q_2 within a particle beam of length ΔL_2 and cross-sectional area A_2 is given by

$$q_2 = n_2(A_2 \, \Delta L_2)q \qquad [15\text{-}17]$$

where n_2 is the number of charged particles per unit volume, $A_2 \, \Delta L_2$ is the volume of the beam, and q is the charge of each particle.

Recall that the current I_2 can be written as [15-4]

$$I_2 = n_2 q v_2 A_2 \qquad [15\text{-}18]$$

where v_2 is the speed of the particles in the beam.

Eliminating $n_2 A_2$ between [15-17] and [15-18], we have

$$q_2 v_2 = I_2 \, \Delta L_2 \qquad [15\text{-}19]$$

That is, a current I_2 over a length ΔL_2 is equivalent to a charge q_2 moving at the speed v_2.

The general relation for the magnetic force is

$$F_m = q_2 v_2 B \sin \phi \qquad [15\text{-}20]$$

Substituting [15-19] into [15-21] then yields

$$F_m = I_2 \, \Delta L_2 \, B \sin \phi \qquad [15\text{-}21]$$

15-8 THE MAGNETIC FORCE ON CURRENT-CARRYING CONDUCTORS

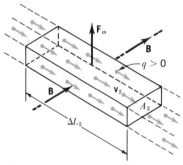

FIGURE 15-24 A beam of positively charged particles moving with velocity v_2 perpendicular to a magnetic field **B**. The beam had a rectangular cross section of area A_2. An element of length ΔL_2 is shown.

The angle ϕ is measured between the direction of **B** and direction of the current in the beam.

The last equation gives the net force on the charged particles in a length ΔL_2 of the beam. The particles have a total charge $q_2 = n_2 q$, and they are moving with velocity $\mathbf{v}_2$. If **B** is perpendicular to $\mathbf{v}_2$, the trajectory of these particles will be a circle, since there are no other net forces acting.[1] But in a wire there *are* other forces: the deflected particles are soon constrained to stay inside the wire by strong forces at the surfaces. As a result, the free electrons in a conductor immersed in a magnetic field of the usual strength do not follow (or, at least, do not complete) circular orbits. No; the moving electrons stay inside and the *current-carrying conductor itself experiences a net transverse force.*

As a matter of fact, if the conductor is of length ΔL_2 and carries current I_2 in an external magnetic field **B,** the magnitude of this transverse force is observed to be given by

[15-22] $F = I_2 \, \Delta L_2 \, B \sin \phi$

That is, *it is exactly the same as the magnitude of the magnetic force on the moving electrons in the beam described above.* This remarkable result is far from obvious. Indeed, if the wire is not held in place, it will be accelerated by this force on it and *gain kinetic energy.* Now a *magnetic* force *never* does work and therefore never imparts kinetic energy. How can this paradox be resolved?

The Hall effect

The basic effect which holds the clue to the answer was first observed by the American physicist E. H. Hall in 1879 and is now called the *Hall effect.* A rigid conductor consists primarily of massive positive ions rigidly held in fixed positions relative to each other by strong intermolecular forces. These ions initially are not moving; only the loosely bound electrons (nearly 2,000 times lighter than the positive ions) move (Fig. 15-25a). As soon as a few electrons are deflected (Fig. 15-25b) by the magnetic field to one surface of the metal, they leave an excess of positive charge (ions) on the opposite surface. As a result, an electric field **E** is established *transverse* to the original direction of electron drift, as shown in Fig. 15-25c. (In 1879, of course, Hall did not know about the existence of electrons anywhere, let alone in a metal! But he knew there was a transverse electric field **E** in the metal wire because in his experiments he observed a transverse electric potential difference ΔV.) This transverse electric field builds up until the electric force on the deflected electrons just balances the transverse force from the magnetic field.

[1] The reason the net electric force from the other charged particles is zero, except near the edges of the beam, is that each particle is surrounded symmetrically by the other particles. Moreover if the particles move very fast, and if the number per unit length is not large, even the effects near the edges are hardly noticeable.

331

+ SECTION 15-8
THE MAGNETIC
FORCE ON
CURRENT-CARRYING
CONDUCTORS

The electric interaction follows Newton's third law: the electrons exert a force back on the charges at the surfaces equal in magnitude to the electric force which these charges exert on the electrons. Thus by showing that the magnitude of this electric force is precisely equal to the magnitude of the magnetic force on the moving electrons we have explained the observed fact that the force on the conductor, [15-22], has the same magnitude as the magnetic force on the beam of particles described in [15-21]. Since the net force **F** involves an electric force which can do work, we can now account for the kinetic energy which can be acquired by the wire. Of course, **F** is a very special net force: it never would exist if the magnetic force had not first deflected the electrons to one side of the conducting wire. For this reason, the force **F** is commonly referred to as a *magnetic force.*

Equation [15-22] gives the force on a *short* straight segment of current-carrying conductor. It may, of course, be used for a long straight conductor provided that the external magnetic field is the same at all points along the conductor. There is, however, no such thing as an infinitely long straight conductor, if only because we must always have a complete loop to circulate charges.

When a conductor's shape is not simple, we can find the resultant magnetic force on it by applying [15-22] to each short segment, taking into account the relative directions of the magnetic field **B** and the current at each point, and then summing, as vectors, the magnetic forces on all segments.

FIGURE 15-25 (*a*) A pictorial representation of the loosely bound electrons in a straight metal wire drifting parallel to the wire in the presence of an electric field $E_\parallel$ but in the absence of an applied magnetic field. The massive positive ions are held rigidly in place by the intermolecular forces. (*b*) As soon as a magnetic field is applied perpendicular to the wire, the loosely bound electrons are deflected to one side of the wire by the transverse magnetic force $F_{m\perp}$. (*c*) In a short time the transverse magnetic force $F_{m\perp}$ is balanced by a transverse electric force $F_{e\perp}$ in the opposite direction. This electric force arises from the accumulation of electrons on one side of the wire and the deficiency of electrons on the opposite side (Hall effect). The positive ions also experience a force from the transverse electric field, and this results in a *net* transverse force on the conductor.

(a) (b) (c)

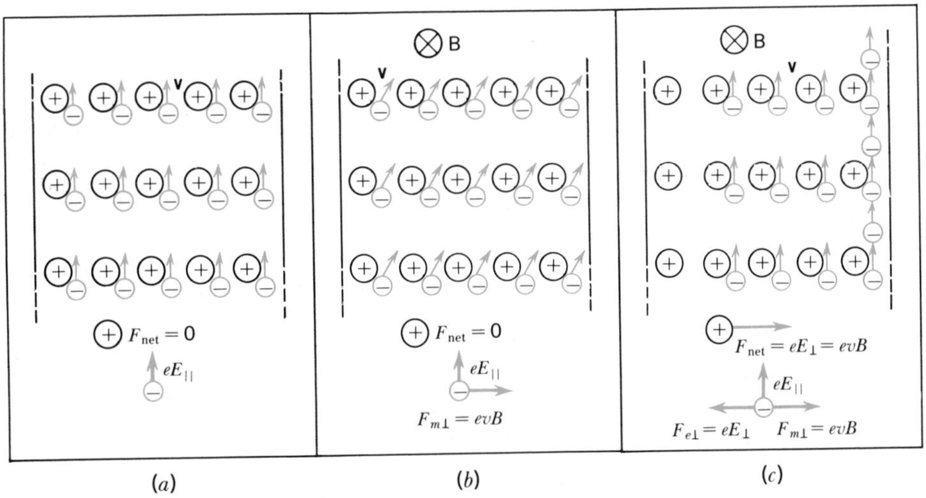

+ 15-9 THE MAGNETIC FORCE
BETWEEN TWO PARALLEL WIRES

Return now to the simplest of all magnetic interactions, that between one long straight current-carrying conductor and a second conductor parallel to the first (see Fig. 15-26). We are now equipped to understand Ampère's experimental observations, by using [15-22] to find the magnetic force between such a pair of conductors in terms of the current I_1 and I_2, the separation distance d, and their common length L.

First, the *directions* of the magnetic forces. As shown in Fig. 15-26, if both currents are in the same direction, the conductor on the right, according to our right-hand rule for magnetic forces, is subject to a magnetic force toward the other conductor. Here we have drawn the field lines $\mathbf{B}_1$ from the left conductor I_1 only, since we wish to find their effect on the right conductor I_2. (A conductor is, of course, subject to no net magnetic force arising from its own magnetic field.) It is equally easy to confirm that the magnetic field of the right conductor acting on the left conductor I_1 will produce a magnetic force toward the right. In short, parallel conductors with currents in the *same* direction attract each other. By similar reasoning if the currents are in opposite directions, the conductors repel each other. Roughly speaking, like currents attract, and unlike currents repel.

Like currents attract; unlike currents repel.

Now to find the *magnitude* of the magnetic force $\mathbf{F}_2$ on conductor I_2. From [15-9] we have, all along I_2,

[15-23]

$$B_1 = k_m \frac{2I_1}{d}$$

The magnetic force on I_2 is, from [15-22]

$$F_2 = I_2 L B_1 = k_m \frac{2I_1 I_2}{d} L \qquad [15\text{-}24]$$

We have taken ϕ to be 90° since the field B_1 is perpendicular to the conductor I_2 at all points. It is convenient to write [15-24] in terms of the force per unit length of conductor

$$\frac{F_2}{L} = k_m \frac{2I_1 I_2}{d} \qquad [15\text{-}25]$$

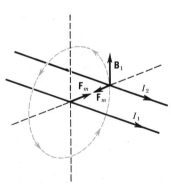

FIGURE 15-26 The attractive magnetic force on a long straight conductor carrying current I_1 in the vicinity of a parallel conductor carrying current I_2 in the same direction. The magnetic field $\mathbf{B}_1$ arises from the current I_1 as shown. Conductor 2 exerts an equal but oppositely directed force on conductor 1.

Notice that the expression on the right of [15-24] is symmetric in the subscripts 1 and 2. This means that the force $\mathbf{F}_1$ on I_1 is equal (and opposite) to $\mathbf{F}_2$. For this particular geometry the magnetic forces do obey Newton's third law.

Suppose that two long parallel conductors each carrying a current of

1 A are separated by 1 m. Then $I_1 = I_2 = 1$ A, $d = 1$ m, and, from [15-25], we have $F/L = 2k_m = 2 \times 10^{-7}$ N/m. The force per unit length on either conductor is exactly 2×10^{-7} N/m. Actually, this relation serves to *define* the ampere as the unit of current. First, recall that the magnetic constant k_m is *assigned* the value of exactly 10^{-7} N/A². The ampere is defined as follows: if two long parallel conductors, separated by 1 m and carring equal currents, interact magnetically with a force per unit length on each of 2×10^{-7} N/m, then *by definition* the current in each is precisely 1 A. The coulomb, the basic unit of charge, is then defined as 1 ampere-second (A-s). And finally the value of the fundamental electric constant, $k_e = 9.0 \times 10^9$ N-m²/C², is found in experiments where the electric force between known charges is measured.

The ampere unit defined experimentally

Next consider the magnetic field of a more complex current configuration, a circular loop of current-carrying wire. To compute the field in detail, even for this rather simple geometry, is a bit messy. We report some of the results of a measurement of this field and then account for these results, at least qualitatively, on the basis of symmetry.

In Fig. 15-27 we show two straight wires that lead current to and from a circular loop of wire. Since the currents for the two *adjacent* leads are in opposite directions, their magnetic fields cancel each other exactly, and we are left with the field of the loop alone. Some of the observed magnetic field lines for such a current loop are shown in Fig. 15-28. Notice that the lines very near the loop are circles. This is exactly what we expect, because near a short piece of the loop, just that piece, and not the more distant portions of the loop, contributes appreciably to the field. Since such a piece is essentially a straight segment, field lines will be circles.

We can also say that the field has rotational symmetry about the axis of the loop. This must be so, since the source current has such symmetry. Therefore, if we know B in one half-plane through the loop axis, we know it everywhere. For example, compare the right half of Fig. 15-28a with the left half of Fig. 15-28a and with the top and bottom halves of Fig. 15-28b.

Near the center of the loop the magnetic field lines are approximately straight (also parallel) and closely spaced. In other words, if you want a reasonably uniform and reasonably strong field, the best place to go is the center of the loop. Even more uniform fields over larger regions of space can be obtained near the center of a set of Helmholtz

+ 15-10 THE MAGNETIC FIELD OF A CURRENT LOOP

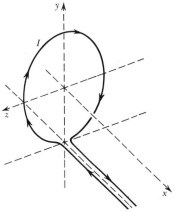

FIGURE 15-27 A circular current loop. The input lead is placed adjacent to the output lead, and there is no net magnetic field from these two leads.

FIGURE 15-28 The observed magnetic field lines from a circular current loop (*a*) in the *xz* plane and (*b*) in the *xy* plane. The *x* axis is perpendicular to the plane of the loop.

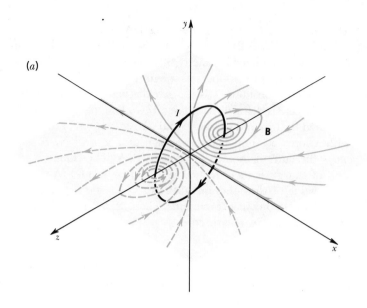

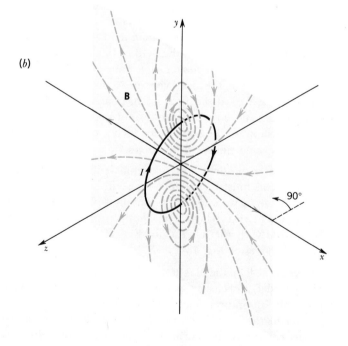

FIGURE 15-29   The observed magnetic field lines in the vicinity of a pair of circular current loops carrying current in the same sense (Helmholtz coils).

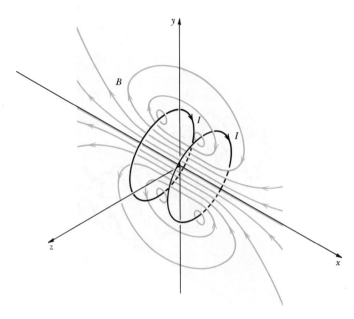

coils (Fig. 15-29) or near the center of a solenoid (Fig. 15-30). The magnitude of the magnetic field at the *center* of a single current loop is (see Problem 15-19)

$$B = 2\pi k_m \frac{I}{a}$$
[15-26]

where I is the current in the loop and a is the radius of the loop. The magnitude of the field at the center of a set of Helmholtz coils is more complicated. The field at the center of a *long* solenoid is given very simply by

$$B = 4\pi k_m nI$$
[15-27]

Here, n is the number of turns of wire per unit length of the solenoid, and I is again the current in the wire. Notice that this field is independent of the length of the coil and of the *total* number of turns.

As remarked earlier, the magnetic field for any geometry is directly proportional to the current. A current I in a circular coil of N turns, all closely spaced and wound in the same sense, looks like a current NI in a single turn. Thus, such a multiple coil, unlike the spread-out

FIGURE 15-30   The observed magnetic field lines in the vicinity of a long current-carrying solenoid.

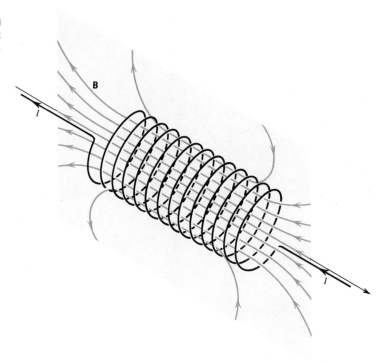

solenoid, does have a magnetic field proportional to the *total* number of turns.

+   **15-11 MAGNETIC FORCE ON A CURRENT LOOP**

In contrast to a straight wire, a loop carrying current in a plane is not acted upon by an unbalanced force when immersed in a uniform magnetic field, but it does experience a net torque, as we shall show in an important special case.

Suppose the loop is rectangular, with two of its sides perpendicular to the field and with the normal to the loop making an angle θ with **B** (see Fig. 15-31). Because the magnetic forces on the sides of length a are equal in magnitude, along the same line, and opposite in direction, they cancel. The magnetic forces $\mathbf{F}_m$ on the sides of length b are also equal in magnitude and opposite in direction, but they are not along the same line in space. Thus the net torque is

Magnetic torque on a current loop

[15-28] $N = 2F_m \dfrac{a}{2} \sin \theta = F_m a \sin \theta$

From [15-22] we have

$$F_m = IbB \qquad\qquad [15\text{-}29]$$

where I is the current in the loop. (Because the magnetic field is perpendicular to the sides of the length b, we took $\phi = 90°$ in [15-22].)

Combining [15-28] and [15-29] yields

$$N = I(ab)B \sin \theta \qquad\qquad [15\text{-}30]$$

It is seen that the loop will be in rotational equilibrium only if $\sin \theta = 0$, that is, only if the normal to the loop is aligned with the magnetic field. The torque, as well as being proportional to I, is proportional to the area ab of the loop. This result holds true for a plane loop of arbitrary shape.

If the current direction were reversed every time the loop passed through the equilibrium position, the loop could be kept spinning continuously. This is the principle of the electric motor.

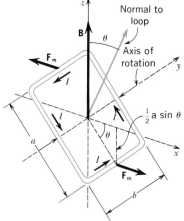

FIGURE 15-31 The magnetic forces $\mathbf{F}_m$ on a pivoted current loop in a magnetic field **B**. Only the forces on the sides of length b parallel to the axle (which is perpendicular to **B**) are effective in rotating the rectangular loop. When the plane of the loop is perpendicular to **B**, there is no net torque; if the loop is at rest in this position, it will remain at rest.

+ **15-12 MAGNETS**

In its behavior when immersed in a magnetic field, a current loop reminds us strongly of a compass needle. The object experiences a torque which tends to align its axis with the field. This resemblance is not accidental. All material magnets, e.g., the compass needle, in effect *are* current loops or series of current loops (solenoids). The currents in question are atomic currents, as indicated in Fig. 15-32a. This model of a magnet was first suggested by Ampère. Along each interior boundary the currents in adjacent loops exactly cancel each other; there is no net current inside the material. But at the outer surface there is no cancellation, and, as Fig. 15-32b shows, a current

Atomic model for a magnetic material

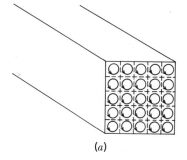

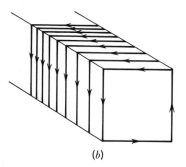

FIGURE 15-32 Model of a permanent magnet showing (a) localized current loops all aligned, with the planes of the loops perpendicular to the lengthwise direction of the magnet and (b) the net current on the surface of the magnet with a distribution similar to that of a long solenoid.

(a)   (b)

runs around the exterior of the material exactly as in the case of the solenoid.

There are two types of atomic electric currents. In a simple atomic model one may think of each electron as orbiting its parent nucleus. For every electron orbiting in the clockwise sense there usually is a nearby electron orbiting in the counterclockwise sense. The net magnetic effect of the revolving electrons is then zero. But in addition to its orbital motion, each electron can be imagined as a little sphere of charge spinning about an internal rotation axis. The spinning charge constitutes a current loop, and it is the magnetism of electron spin which is responsible for the properties of iron and similar materials. The rate at which an electron spins about its own axis is not affected by any external influence. One might say that spinning charged particles are the only true permanent magnets in nature. In strongly magnetic materials, such as iron, electron spins are not paired off in opposite rotation senses, as in most materials but are aligned, (Fig. 15-32a), and their magnetic effects do not cancel.

SUMMARY   A particle of charge q moving with velocity **v** through a magnetic field **B** experiences a magnetic force $\mathbf{F}_m$ having the magnitude

[15-11] $F_m = qvB \sin \phi$

where ϕ is the angle between **v** and **B**. The line of orientation of this force is perpendicular to the plane containing **v** and **B,** and the direction (sense) along this line is given by the right-hand rule for magnetic forces.

Similarly a segment of wire of length ΔL carrying a current I through a magnetic field **B** experiences a force **F** having magnitude

[15-22] $F = I \Delta L\, B \sin \phi$

where ϕ is now the angle between the direction of the current and the magnetic field. The direction of this force is given by the right-hand rule, with the direction of the current replacing the direction of the velocity **v.**

If in a time Δt a net charge Δq is transported through an area A, the current through A is given by

[15-3] $I = \dfrac{\Delta q}{\Delta t}$

For many common (Ohmic) devices, the current I through the device is directly proportional to the electric potential difference across it (Ohm's law)

$$\Delta V = RI \qquad [15\text{-}5]$$

The magnetic field **B** a radial distance d away from a long straight conductor carrying current I has a magnitude

$$B = k_m \frac{2I}{d} \qquad [15\text{-}9]$$

and the line of orientation of **B** is the line along which a test charge can move and experience no force. This unique line of zero magnetic force lies in a plane perpendicular to the wire and is tangent to a circle of radius d centered on the wire. The direction (sense) of **B** along this tangent is given by the right-hand rule for magnetic fields.

Consider a simple arrangement in which two particles, each with charge q and velocity **v,** interact with each other both by the electric force and the magnetic force. We suppose that the two particles are traveling side by side separated by a distance r, as shown in Fig. 15-33. If both charges are of the same sign, e.g., both positive, they repel each other by an electric force whose magnitude, from [13-2], is

$$F_e = \frac{k_e q^2}{r^2}$$

The particles also interact magnetically. The upper charge in Fig. 15-33 creates a magnetic field which is into the paper at the location of the lower charge. Thus, the magnetic force on the lower charge is upward. In the same fashion it is easy to see that the magnetic force on the upper charge is downward. In other words, the two particles attract each other magnetically, and in this instance the magnetic force *is* along the line connecting the two particles. The magnitude of the magnetic force **F**$_m$ is given by [15-1] with both θ and ϕ equal to 90°

$$F_m = \frac{k_m q^2 v^2}{r^2}$$

The particles repel electrically but attract magnetically. What is the ratio of the two forces? Combining these two equations, we find that

ADDENDUM

**+   THE MAGNETIC FORCE
AND REFERENCE FRAMES**

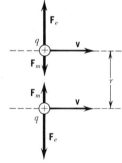

FIGURE 15-33   The magnetic forces experienced by two particles each with the same charge and moving parallel to the other with the same speed. The magnetic forces are attractive, and the electric forces are repulsive, but all lie along the line connecting the two particles.

[15-31]
$$\frac{F_m}{F_e} = \frac{k_m q^2 v^2 / r^2}{k_e q^2 / r^2} = \frac{k_m}{k_e} v^2$$

The charge magnitudes and the separation distance cancel, and we see that the force ratio depends only on the common speed of the two particles and k_m/k_e, the ratio of the fundamental magnetic and electric interaction constants. This ratio has the value

$$\frac{k_m}{k_e} = \frac{10^{-7}\ N/A^2}{9.0 \times 10^9\ N\text{-}m^2/C^2}$$

or, since 1 A = 1 C/s

[15-32]
$$\frac{k_m}{k_e} = \frac{1}{9.0 \times 10^{16}\ m^2/s^2} = \frac{1}{(3.0 \times 10^8\ m/s)^2}$$

The quantity 3.0×10^8 m/s is a speed. Indeed, it is a very special speed, that of light through empty space. Designating it by its customary symbol c, we have

[15-33]
$$\frac{F_m}{F_e} = \left(\frac{v}{c}\right)^2$$

The magnetic force is smaller than the electric force by the square of the speed ratio v/c. Unless the particles move at relatively high speeds, the magnetic attraction is altogether trivial as compared to the electric repulsion. For example, if both particles moved at 18,600 miles/s (one-tenth the speed of light), [15-33] shows that the electric force would be 100 times greater than the magnetic force.

The fact that the ratio k_m/k_e turns out to be just $1/c^2$ suggests all sorts of fascinating things about electromagnetism and about light. For example, why does the speed of light appear in a fundamental relation between electricity and magnetism? Is light in some sense an electromagnetic phenomenon? (We see in Chapter 20 that it is.) *What* is so special about the speed of light? (The speed of light is the fundamental constant of the theory of relativity and might be called the speed limit of the universe.)

The magnetic force between a pair of particles is only small in comparison to the electric force but can actually be made *zero*. Again we regard the two particles moving side by side at the same speed, but this time we do so as an observer in a reference frame also moving to the right at the speed v. In this frame the particles are momentarily at rest, and so the magnetic force is zero. Only the electric force remains. We have turned off the magnetic force by choosing a different reference frame!

Turning a magnetic force off by changing reference frames

In viewing magnetism from an increasingly fundamental standpoint, we began with a pair of magnets and saw that the magnetic force is manifested, even without magnets, when two current-carrying conductors interact. At a still more basic level, we saw the magnetic interaction between a pair of charged particles competing (weakly) with the ever present electric interaction. Now, merely by the proper choice of a reference frame, we have removed the magnetic effect entirely. Not only does electromagnetism have *fundamentally* nothing to do with magnets as such, electromagnetism has, it appears, nothing to do with magnetism. Let us reverse our consideration of the two charges, first viewing them in a reference frame in which they are at rest. Now there is an electric force but no magnetic force. If we transform to another reference frame, one in motion relative to the charged particles, the magnetic force between them is created. Clearly, the existence of magnetism is intimately bound up with the matter of reference frames and the speed of light. Indeed, electromagnetism is very closely related to the special theory of relativity. As we shall see in Section 21-7 the magnetic force exists *because* of some very general requirements of relativity theory.

Magnetic effects are due basically to electric effects and relativity.

PROBLEMS

15-1 A current-carrying wire has a cross-sectional area of 1 mm². What is the drift speed of the electrons if the wire contains 10^{23} free electrons per cubic centimeter (about one free electron per atom in a metal) and produces a current of 200 A?

15-2 (*a*) What is the number of conduction electrons per unit volume in a metal conductor 1 mm in diameter carrying a current of 10^{-3} A (1 mA) if the drift speed of these electrons is 10^{-7} m/s? (*b*) What is the number of negative ions per unit volume in a cylindrical tube of water 1 cm in diameter carrying a net current of 1 mA in a direction parallel to the axis of the tube if the speed of the negative ions is 3.5 cm/s while the same number of positive ions per unit volume move in the opposite direction at 2.0 cm/s?

I_1 (into paper) — 40 A — 0.20 m — 0.40 m — I_2
PROBLEM 15-5

15-3 (*a*) What is the resistance of an Ohmic conductor (load) connected across a battery if there is a current of 3 A in the conductor when the electric potential difference across the conductor is 6 V? (*b*) How much electronic kinetic energy is lost per second in this conductor? *Hint:* How much charge drifts across the electric potential difference in 1 s?

15-4 What magnetic field **B** is produced 4.00 cm away from a long straight wire carrying a current of 200 A?

15-5 Two long parallel wires shown in the figure are 0.60 m apart. If the left wire carries a current of 40 A into the paper, what must be the direction and magnitude of the current in the right wire if the magnetic field is to be zero at the point between the wires 0.20 m from the left wire?

15-6   Three long straight wires in the xz plane are all parallel to the z axis. The first wire, along the line $x = -20$ cm, has a current of 200 A in the direction of the positive z axis. The second and third wires, along the lines $x = -10$ cm and $x = +10$ cm respectively, have currents of 100 A in the direction of the negative z axis. What are the direction and magnitude of the magnetic field along the line $x = 0$ in the xz plane?

15-7   The uniform magnetic field in a research magnet has a magnitude 1.5 T and a horizontal northward direction. What are the magnitude and direction of the magnetic force on an electron moving with a speed 3×10^7 m/s in the following directions: (a) north, (b) south, (c) east, (d) west, (e) in a horizontal plane in a northeasterly direction, 53° north of east, (f) vertically upward, and (g) vertically downward?

15-8   A proton moves in the direction of $+y$ axis with a speed of 3.0×10^3 m/s. A uniform electric field of magnitude 6.0×10^3 V/m exists in the $+z$ direction. What is the magnitude and direction of the magnetic field **B** necessary to keep the proton moving in a straight line?

15-9   A rigid rectangular coil in a vertical east-west plane hangs with its lower end in the gap of the magnet described in Problem 15-7. The coil is suspended from one arm of an equal-arm balance and a pan having the same mass as the coil hangs from the other end. The lower end of the coil is 5 cm wide, and the magnetic field may be assumed to be zero outside the magnet gap. (a) If the equal-arm balance is to be "balanced" by adding weights to the pan when there is a current in the coil, what must be the direction of this current in the lower end of the coil? (b) If this current is 1 A, what weight must be added to the pan to attain balance?

Pole force of magnet

—5cm

B northward
1.5 T

East

PROBLEM 15-9

15-10   What is the speed of a beam of electrons when the simultaneous influence of an electric field of 2.6×10^5 V/m and a magnetic field of 0.4 T (both fields perpendicular to the beam) produces no deflection of the beam? Show the relative directions of **v, B, E, F**$_e$, and **F**$_m$ in a diagram where the direction of the *electron* motion is into the paper and **B** is to the right.

15-11   A deuteron (same charge as a proton but about twice its mass) travels in a circle of radius 40 cm in a uniform magnetic field of 1.5 T. (a) Find the speed of the deuteron. (b) Find the time required for it to make 1 rev. (c) Through what electric potential difference would the deuteron have to be accelerated in order to gain this speed?

15-12   In 1897 J. J. Thomson used the following procedure to determine the charge-to-mass ratio for the electron. He first measured E, L, and y_L in the experiment described in Example 13-11 (only he found that his particles were negatively charged). Then he applied a magnetic field perpendicular to the paper in Fig. 13-18 and increased the magnitude B of this magnetic field until y_L became zero. (a) Show how he could determine q/m as soon as he measured this final value of B. (b) Does the **B** have to be directed into the paper or out of the paper in Fig. 13-18?

15-13 The magnetic field of the Earth at the equator has a horizontal northward direction, and its magnitude is 0.70 G (0.70×10^{-4} T). (a) In order to have a proton encircle the equator with constant speed should the proton be fired to the east or to the west? (Assume that the atmosphere has no retarding effect, or assume that a large doughnut-shaped vacuum tube has been constructed around the equator.) (b) With what speed must the proton be fired? (c) Compare this magnetic force with the gravitational force on the proton.

15-14 A cyclotron accelerates protons to a final energy of 8.0 MeV. The magnetic field is 1.6 T. (a) At what frequency must the electric potential difference across the D's be cycled? (b) What minimum radius is required for the D's if this final energy is to be achieved? (c) If the maximum electric potential difference applied to the two D's is 100 kilovolts (kV), how many cycles must the protons complete before achieving the final energy?

15-15 What is the force per unit length on two long straight wires 4.00 cm apart when each carries a current of 200 A?

15-16 What must be the direction of a magnetic field if it is to cause the loop shown in the figure to rotate counterclockwise as seen from above?

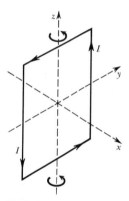

PROBLEM 15-16

15-17 (a) If an electron in a hydrogen atom is thought of as revolving around the proton in a circular orbit of radius 5×10^{-11} m at a constant speed 2×10^6 m/s, what is the current associated with this electron? (b) What is the magnetic field at the proton due to this current?

15-18 (a) Starting with a general relationship describing a small element of a beam of moving charges in terms of a small element of current (see Section 15-8), show that the magnitude of the magnetic field from an element of current is $B = (k_m I_1 \, \Delta L_1 \sin \theta)/r^2$. Hint: Also look at Section 15-5. (b) What is the direction of this magnetic field from an element of current? Hint: Look at Fig. 15-14.

c 15-19 (a) Using integral calculus and the result of Problem 15-18, show that the magnitude of the magnetic field at the center of a circular current loop is $B = 2\pi k_m I/a$, where I is the current in the loop and a is the radius of the loop. (b) What is the direction of the magnetic field at the center?

15-20 The restoring torque of a twisted fiber is directly proportional to the angular displacement, or twist. Explain how a current-carrying loop immersed in a uniform magnetic field can be used as a current-measuring device (such an instrument is called a galvanometer).

CHANGING ELECTRIC AND MAGNETIC FIELDS

We know that a particle of charge **q** located in an electric field **E** experiences a force of magnitude $\mathbf{F}_e = q\mathbf{E}$. Moreover, if this particle is moving with velocity **v** through magnetic field **B**, it experiences an additional force of magnitude $F_m = qvB \sin \phi$, where ϕ is the angle between **v** and **B**. The fields **E** and **B** are given respectively by Coulomb's law [13-5] and its magnetic counterpart [15-10]. In general these fields may vary from point to point. These formulas therefore give the electric and magnetic forces as functions of position.

What happens to a charged particle immersed in an electric or magnetic field which changes *with time?* Is it still a matter of using the same expressions for the forces, merely taking into account that **E** and **B** vary with time as well as with position? After all, this is what we do in mechanics. For example, in dealing with a satellite receding from the Earth, we must take into account the fact that the gravitational force falls off with distance; therefore we say the force on the receding satellite decreases with time.

But the electric and magnetic interactions are different. When the magnetic field changes with time, a complete specification of a charged particle's motion requires $F_m = qvB \sin \phi$ and *an additional force.* The additional force is not magnetic; it is actually an electric force, an induced electric force. Similarly, when the electric field changes with time, there comes into play a force *additional to* $\mathbf{F}_e = q\mathbf{E}$. And, curiously, this new force is not an electric force but an induced magnetic force. Let us now look into these remarkable cross relationships.

16-1 MAGNETICALLY INDUCED CURRENTS

After 1820, when Oersted discovered that an electric current produces magnetism, it was natural to ask whether the reverse effect exists, i.e., whether magnetism might produce current in a wire. For example, the current in the circular loop of wire shown in Fig. 16-1a creates a magnetic field. The direction of the field at the center of the loop

FIGURE 16-1 (*a*) A current in a circular wire loop produces a magnetic field at the center of the loop in a direction perpendicular to the plane of the loop. (*b*) But a magnetic field at the center of a circular wire loop in a direction perpendicular to the plane of the loop does not necessarily mean that a current exists in the loop.

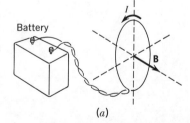
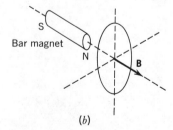

(*a*)   (*b*)

is shown in the figure by the straight arrow, and the sense of the current is shown by the curved arrow. What about the arrangement shown in Fig. 16-1b? Here we have a magnetic field (again shown by a straight arrow) arising from a magnet. In this field a circular loop of wire has been placed with the same orientation relative to the field at its center as the loop of Fig. 16-1a. Is there a current in the second loop?

Faraday shared, to use his own words, "the hope of obtaining electricity from ordinary magnetism." Indeed, in 1831 he performed the very experiment portrayed in Fig. 16-1b. It failed in the sense that no current was found with the magnet *fixed* in place, but a current was induced in a way that Faraday had not anticipated. If he *moved* the magnet, or if the magnetic field at the loop otherwise varied with time, a current *was* produced in the conducting loop.

The current induced in a loop can transfer energy: it can heat the filament of a lamp (Fig. 16-2). How does a moving magnet (or, more generally, a changing magnetic field) generate energy in an electric circuit? At this point it would be enormously complicated to try to answer this question for all the possible variations of Faraday's experiment. Happily, phenomena related to induced current and the transfer of energy can be described rather simply in terms of two new concepts, *electromotance* and *magnetic flux*. After taking time to explain these ideas we shall return to the induction phenomena themselves.

The Faraday effect

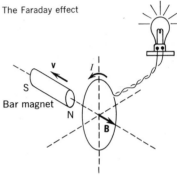

Bar magnet

FIGURE 16-2 A *changing* magnetic field at the center of a circular wire loop in a direction perpendicular to the plane of the loop produces a current in the wire loop. This induced current can be used to light an electric bulb.

To start on familiar ground, let us see how energy is transferred, not in an electrical circuit, but in a mechanical circuit.

16-2 ELECTROMOTANCE

Consider first the simple roller coaster of Fig. 16-3. A marble placed at the highest point and then released from rest rolls downhill and uphill until it completes the circuit and finally returns to its starting point, where it is once more at rest. In terms of energy, we can say that the Earth-marble system loses gravitational potential energy and gains kinetic energy in an equal amount as the marble goes downhill and conversely for the uphill portions of the path, so that the total energy of the system remains constant. Said differently, the force of gravity does positive work on the marble as it speeds up in going downhill and negative work on the marble as it slows down going uphill, and the total work done by the gravitational force over the round trip is zero. For this reason we characterize the gravitational force as a conservative force.

A mechanical model illustrating the transfer of energy in an electric circuit

Work done by the gravitational force over a round trip is zero.

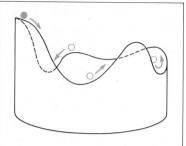

FIGURE 16-3 A simple marble roller coaster. The marble moves on a frictionless track, losing kinetic energy on the uphill portions of its trip but regaining it on the downhill portions.

Nonconservative forces dissipate energy in collisions.

The player supplies the needed energy through a nonconservative force.

Power defined: time rate of doing work

Power unit, the watt (W)

The electrical-mechanical analog

Now consider the mechanical circuit shown in Fig. 16-4. In this toy, marbles are dropped into the hole at the top. Each marble rolls along the uppermost inclined ramp until it makes an inelastic collision, a collision in which it *loses* essentially all its kinetic energy, with the sidewall. Then the marble, starting again from rest, falls through a hole to the adjoining ramp, and the process repeats: the marble gains kinetic energy while rolling downward, but this newly acquired energy is lost in the next collision. The outcome of this situation is that the marble ends up at the bottom with the same kinetic energy it had when it started at the top: *none*. Unlike the roller coaster, the marble cannot make a round trip to the starting point by itself.

The system has lost gravitational potential energy without a net gain in kinetic energy. We know from Chapter 11 what happens: on each ramp some potential energy is transformed into ordered kinetic energy, which, in turn, is changed into disordered thermal energy by the nonconservative dissipative forces that act during the collision. Of course, the toy does not become so hot that it glows like an electric light; nevertheless, the principle is similar in that the light from the hot filament of an electric bulb has its origin in the inelastic collisions between the electrons in the filament and the other particles of the material.

The ultimate source of energy to bring the marble back to its starting point is the player. In increasing the Earth-marble separation, i.e., lifting the marble to the top of the toy, he does work, and this work shows up as a potential-energy increase. In summary, then, the player performs work on the system along one segment of the marbles' closed path, increasing the potential energy of the system. But in another segment of the path this potential energy is transformed into thermal energy. All that remains of the initial work against the gravitational force is "hotter" marbles and walls.

Note that energy is fed into the system by a *nonconservative* force. Unlike gravity, which does zero net work around a marble's closed circuit, the force of the player's hand performs positive net work during each round trip. The more trips made, the more energy introduced into the system. Thus the player, at least until he tires, functions as a *source of power,* where power means simply *energy or work per unit time.* Power is measured in units of joules per second, or watts (W).

The electrical counterpart of our marble game is easily recognized. It is a simple circuit, driven by a battery. The analogy is detailed in Table 16-1. In the electrical system, the energy supplied by the battery

TABLE 16-1   Corresponding Quantities

Marble Circuit (Fig. 16-4)	Electric Circuit (Fig. 16-5)
1 Person lifting marbles	1 The battery
2 Person eating	2 Charging the battery
3 Nonconservative force of hand	3 Nonconservative chemical forces within the battery
4 Marbles	4 Conduction electrons
5 Conservative gravitational force	5 Conservative electric force
6 Gravitational potential energy	6 Electric potential energy
7 Nonconservative dissipative forces during inelastic collisions along ramp	7 Nonconservative dissipative forces during inelastic collisions in resistor

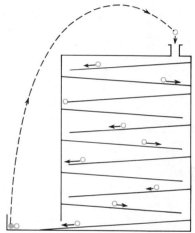

finally appears as thermal energy in a resistive load, say the filament of a bulb in another segment of the circuit (see Fig. 16-5). (The conventional symbol for a resistor bears an amusing resemblance to the zigzag series of ramps in Fig. 16-4.) Again there is a conservative force in the system, the Coulomb electric force arising from electric charges on the plates of the battery and in the conducting wires. And again there is a nonconservative force, that which arises from chemical reactions in the battery. The battery does net work on the charges each time they traverse it; the conservative Coulomb force does no net work around the closed circuit. Hence the battery is the power source, now electrical instead of mechanical.

Looking back at Faraday's experiment (Fig. 16-2), we can appreciate that ordinary Coulomb forces in the conductor cannot be responsible for heating the filament. There must exist a nonconservative force, somehow associated with the changing magnetic field, which supplies net energy to the circuit.

Because such nonconservative forces, whether arising in a changing magnetic field or in an ordinary chemical battery, act only on charged particles, we still call them electric forces, but in order to

FIGURE 16-4   A marble game in which the kinetic energy gained by a marble during the downhill portions of the trip is converted to disordered thermal energy during collisions with the walls. Therefore an external force (from a hand) must do work against the force of gravity in order to return the marble to its starting point at the top of the track.

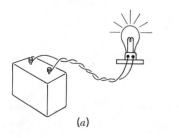

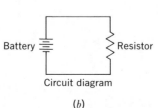

Battery ᆖ   ⌇ Resistor

Circuit diagram

(a)   (b)

FIGURE 16-5   An electrical circuit (a) and associated diagram (b). The energy supplied by the battery finally appears as disordered thermal energy in the heated filament (a wire of large resistance) of the light bulb.

distinguish them from the conservative Coulomb force we designate them by a special symbol. Retaining the symbol **E** for the force per unit charge arising from the Coulomb interaction, we use the symbol **E'** for the force per unit charge associated with the nonconservative forces.

To discuss energy changes produced by **E'** we cannot set up a potential energy or potential; that is possible only for a conservative field. But we can do the next best thing. Let W be the work done by E' on a particle of positive charge q as the particle moves over a particular path between two points in space. Then the magnitude of the work per unit positive charge will be called the *electromotance* $\mathcal{E}$ over that path. Thus, when **E'** is *uniform* along the path, the work per unit positive charge is simply (see [8-13])

Electromotance: work per unit positive charge

$$\mathcal{E} = \frac{W}{q} = \frac{F \cos \theta \, \Delta s}{q} = E' \cos \theta \, \Delta s = E'_{\parallel} \Delta s \qquad [16\text{-}1a]$$

where θ is the angle ($< 180°$) between **E'** and the displacement $\Delta\mathbf{s}$. Here $E'_{\parallel}$ is the component of **E'** in the direction of the total displacement $\Delta\mathbf{s}$. But, *in general*, we must follow our usual procedure (Section 8-6) for computing the work done by a *nonuniform* force, and we obtain

$$\mathcal{E} = \Sigma E'_{\parallel} \delta s \qquad [16\text{-}1b]$$

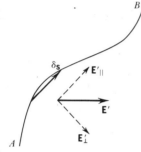

FIGURE 16-6 The components of the force per unit charge **E'** associated with a nonconservative electric force. Only the component of **E'** in the direction of the displacement δs contributes to the work per unit positive charge $\mathcal{E}$, called the electromotance.

where, as shown in Fig. 16-6, $E'_{\parallel}$ is now the component of the force per unit positive charge in the direction of the small element of displacement $\delta\mathbf{s}$ having a magnitude nearly equal to that of the length of the path element.

It is seen from [16-1a] that the electromotance $\mathcal{E}$ and the electric potential difference ΔV have the same units (volts). Since **E'** is nonconservative, the electromotance generally depends on the particular path followed, whereas the electric potential difference is always independent of path.

[1] Often the historically sanctioned but misleading term *electromotive force* (emf) is used instead of electromotance.

In the examples that follow the concept of electromotance will be applied to find power relations for electric circuits.[1]

EXAMPLE 16-1   A current I exists in a circuit having a source of power with an electromotance $\mathcal{E}$. Show that the power P supplied by the source is equal to $\mathcal{E}I$.

SOLUTION   If charge ΔQ moves through the source in a time Δt, then the current is $I = \Delta Q/\Delta t$. Since the electromotance is simply the work done per unit charge, the amount of work done on the charge ΔQ is $\mathcal{E} \Delta Q$. But the power is by definition the work per unit time

$$P_{\mathcal{E}} = \frac{W}{\Delta t} = \mathcal{E} \frac{\Delta Q}{\Delta t} = \mathcal{E} I \qquad [16\text{-}2]$$

In words, the power supplied by the source is the product of the electromotance and the current. For example, a 6-V battery carrying a current of 0.5 A supplies 3 W of power.

A battery can either supply power (discharge) or receive power (charge). When it delivers power, work is done by the battery on the charges which constitute the current in the circuit. The current is then in the same direction as the nonconservative electric field $\mathbf{E}'$. In the charging process, a second power source pushes current backward (counter to $\mathbf{E}'$) through the battery, thereby supplying energy to it.

EXAMPLE 16-2

Ohm's law

Certain components in electric circuits, e.g., ordinary carbon or wire resistors at constant temperature, obey Ohm's law: the electric potential difference ΔV between the two terminals of the component is directly proportional to the current I in the component, the constant of proportionality being called the *resistance R*. In symbols

$$\Delta V = RI$$

Show that the energy per unit time dissipated as heat in such an ohmic component is equal to the square of the current times the resistance

$$P_R = I^2 R \qquad [16\text{-}3]$$

The work done by the Coulomb electric field as a unit charge moves through the component from one terminal to the other is simply the electric potential difference ΔV. Thus if a charge ΔQ moves through the component in a time Δt, this work is $\Delta V \Delta Q$. This work results in a momentary increase in the ordered kinetic energy of the moving charge, but this ordered energy is soon dissipated into disordered thermal energy as the result of collisions. Thus, the energy dissipated per unit time, which is the power P_R dissipated in the Ohmic component, is

$$P_R = \Delta V \frac{\Delta Q}{\Delta T} = (\Delta V)I = I^2 R$$

A 10-Ω resistor carrying 2 A dissipates energy at the rate $P = I^2 R = (2 \text{ A})^2 (10 \ \Omega) = 40 \text{ W}$.

EXAMPLE 16-3

The two terminals of an Ohmic component having resistance R are connected by perfectly conducting wires to the two terminals of a power supply having an electromotance $\mathcal{E}$. The internal resistance of the power supply, representing energy dissipation within it, is r. Show that current in this circuit is

[16-4] $$I = \frac{\mathcal{E}}{R + r}$$

SOLUTION

By energy conservation we know that the power supplied [16-2] by the source must equal the power dissipated [16-3] in the load (resistance R) and in the internal resistance r of the source

$$P = P_R + P_r$$
$$\mathcal{E}I = I^2R + I^2r$$
$$I = \frac{\mathcal{E}}{R + r}$$

EXAMPLE 16-4

Resistors in parallel

The perfectly conducting leads of a 1-Ω resistor and a 2-Ω resistor are tied together at both ends, and this *parallel* combination is connected to the two terminals of a power supply as shown in the circuit diagram (Fig. 16-7). The electromotance of the power supply is 6 V, and its internal resistance is $r = \frac{1}{3}$ Ω. (*a*) What is the ratio of the current in the 1-Ω resistor to the current in the 2-Ω resistor? (*b*) What is the current in the power source?

SOLUTION

Let I_1, I_2, and I_0 be the current in the 1- and 2-Ω resistors and the power source, respectively. The law of electric-charge conservation demands that in a steady state whatever charge comes out the source must go through the load before it goes back into the source. Hence,

$$I_1 + I_2 = I_0$$

We also know that the two load resistors have the same electric potential difference ΔV from one end to the other. Therefore

$$R_1I_1 = R_2I_2$$

or

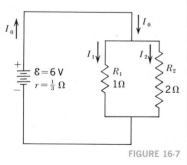

FIGURE 16-7

$$\frac{I_1}{I_2} = \frac{R_2}{R_1} = \frac{2}{1}$$

There is twice as much current in the 1-Ω resistor as in the 2-Ω resistor. Indeed, the parallel combination is equivalent to a single resistance R carrying the total current

$$I_0 = I_1 + I_2$$

or

$$\frac{\Delta V}{R} = \frac{\Delta V}{R_1} + \frac{\Delta V}{R_2}$$

so that

$$\frac{1}{R} = \frac{1}{R_1} + \frac{1}{R_2} \qquad [16\text{-}5]$$

or

$$R = \tfrac{2}{3}\Omega$$

Therefore, using the result of Example 16-3,

$$I_0 = \frac{\mathcal{E}}{R + r} = \frac{6}{\tfrac{2}{3} + \tfrac{1}{3}} = 6 \text{ A}$$

Query Can you show that the equivalent resistance R of two separate resistances R_1 and R_2 connected *in series* is $R = R_1 + R_2$?

Use the energy-conservation principle to determine directly the electromotance $\mathcal{E}$ of the source in the circuit shown in Fig. 16-8.

EXAMPLE 16-5

The energy-conservation principle states that in a steady state

SOLUTION

Power supplied to circuit = power dissipated in circuit

Using the results of Examples 16-1 and 16-2, we can write the power supplied by the source as the product of the electromotance $\mathcal{E}$ and the current I, and we can write the power dissipated in a resistance R as RI^2. Therefore,

$$\mathcal{E}I = R_1 I_1^2 + R_2 I_2^2 + R_3 I_3^2 + rI_{\mathcal{E}}^2,$$

in which the fourth term on the right accounts for the dissipation within the power supply itself. Thus,

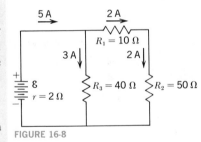

FIGURE 16-8

$$\mathcal{E}(5 \text{ A}) = (10 \ \Omega)(2 \text{ A})^2 + (50 \ \Omega)(2 \text{ A})^2 + (40 \ \Omega)(3 \text{ A})^2 + (2 \ \Omega)(5 \text{ A})^2$$

or

$$\mathcal{E} = 130 \text{ V}$$

How is the induced current in the Faraday effect (or, more basically, the electromotance which drives that current) related to the changing magnetic field? It turns out that what actually matters is not the changing magnetic *field* but the changing magnetic *flux*. Before

16-3 MAGNETIC FLUX

defining this term, let us verify that a changing magnetic field per se does not have to lead to an induced current.

Suppose we have a long straight current-carrying conductor surrounded by its magnetic field, together with a circular conducting loop which can be put in various locations and orientations relative to the straight conductor (see Fig. 16-9). When the current I_1 in the straight wire changes with time, so does the magnitude of the magnetic field it creates. A current I_2 in the circular loop indicates that an electromotance has been created around this loop. In Fig. 16-9a we have the loop in the yz plane. (For the direction of I_1 shown, the magnetic field through the loop is in the negative x direction.) For this arrangement an induced current I_2 is detected whenever the magnetic field changes through an increase or decrease in I_1. But now consider the situation portrayed in Fig. 16-9b. The loop is still in the yz plane, but the straight conductor now passes directly through its plane as a diameter. As the figure shows, the magnetic field in the right half of the loop is directly opposite to that in the left half. As before, when current I_1 varies, so does the magnitude of the magnetic field; but now there is *no* induced current. The same applies for the arrangement in Fig. 16-9c. Here the loop has been turned to lie in the xy plane, so that the straight conductor is perpendicular to the plane of the loop. Again the changing magnetic field gives rise to *no* induced current.

The changing quantity which *is* directly related to the magnitude of the induced current is the *magnetic flux*. The computation of the magnetic flux involves first the choice of some loop, an actual conducting loop for a real experimental arrangement or merely an imaginary loop in space. In either case, the magnetic flux Φ_B through the chosen loop is defined in general as the product of A, the area of any surface bounded by the loop, and $\langle B_\perp \rangle_A$, the component of the magnetic field perpendicular to this surface averaged over the

Magnetic flux defined

FIGURE 16-9 When the current I_1 in the long straight wire and its magnetic field B_1 change, there is an induced current in the circular loop if the positions of the wire and loop are as shown in (a), but there is no induced current in the circular loop if the positions are as shown in (b) or (c).

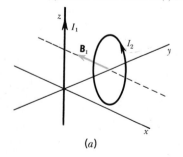

(a)

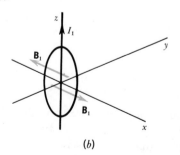

(b)

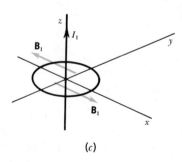

(c)

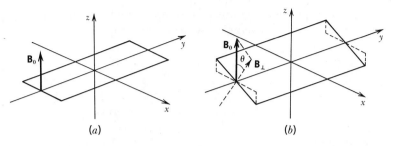

(a)   (b)   (c)

FIGURE 16-10 A uniform magnetic field B_0 is everywhere parallel to the z axis and in the positive z direction. The magnetic flux through a rectangular loop has its maximum magnitude when the plane of the loop is perpendicular to the magnetic field (a), and the magnetic flux is zero when the magnetic field is parallel to the plane of the loop (c). For intermediate orientations (b), the magnitude of the magnetic flux is proportional to the cosine of the angle θ between the magnetic field and the normal to the plane of the loop.

surface of area A. If **B** is *constant over the surface*, this general definition reduces simply to

$$\Phi_B = B_\perp A \qquad [16\text{-}6a]$$

But, *in general*, the flux is given by

$$\Phi_B = \langle B_\perp \rangle_A A \qquad [16\text{-}6b]$$

In [16-8] we shall obtain a precise expression for calculating this average of the perpendicular component. But first, some simpler questions.

The unit of magnetic flux is, of course, the unit of magnetic field times the unit of area. In the mks system, the magnetic field is sometimes assigned the unit webers per square meter. We now understand why: with this assignment the equally significant quantity magnetic flux comes out in the simple unit webers (Wb).

Magnetic-flux unit, the weber (Wb)

Several values of magnetic flux occur in Fig. 16-10. The magnetic field **B** is uniform in the $+z$ direction; it has the constant magnitude B_0. The loops in the three parts all have the same rectangular shape and area A, but they differ in orientation. In Fig. 16-10a, **B** is perpendicular to the loop, and the flux is just $\Phi_B = B_0 A$. But in Fig. 16-10b, the uniform field makes an angle θ with the normal to the loop, so that $B_\perp = B_0 \cos \theta$. The flux is $\Phi_B = B_\perp A = (B_0 \cos \theta) A$. In Fig. 16-10c the angle is $\theta = 90°$; there is no normal component of **B,** and the flux is zero.

If the magnetic field is not uniform over the surface of area A, we can compute Φ_B by imagining the surface divided into a number of parts of area δA, each one small enough so that the magnetic field over it is constant. Then the total flux is just the sum of contributions from each part

Magnetic flux for a nonuniform magnetic field

$$\Phi_B = \Sigma B_\perp \, \delta A \qquad [16\text{-}7]$$

The above expression is equivalent to the original definition [16-6b]; the very meaning of averaging with respect to area is

$$[16\text{-}8] \qquad \langle B_\perp \rangle_A = \frac{1}{A} \Sigma\, B_\perp\, \delta A$$

In the application of [16-7] and [16-8] care should be taken consistently to use either the outer or the inner normal to the surface at every small patch δA. (We can always imagine the surface as forming part of the boundary of a solid volume; the normal pointing out of the volume is commonly used as the positive normal.) Depending on the local direction of **B**, $B_\perp$ may be positive or negative, so that the right-hand sides of [16-7] and [16-8] are algebraic sums. For instance, if we choose the positive x direction as the direction of the normal at each point of the area bounded by the conducting loop in Fig. 16-9b, then $B_\perp < 0$ for $y > 0$ and $B_\perp > 0$ for $y < 0$. The positive and negative terms $B_\perp\, \delta A$ cancel in pairs and $\Phi_B = 0$. However it is not because Φ_B equals zero that no current is induced in the loop, but rather because the flux is *constant*.

EXAMPLE 16-6   The golf bag shown in Fig. 16-11, a right-circular cylinder of cross-sectional area A, is located in a uniform magnetic field **B**$_0$ parallel to the symmetry axis of the cylinder. (*a*) What is the magnetic flux through the surface bounded by the top rim and consisting of the curved side and the bottom as shown in Fig. 16-11a? (*b*) What is the magnetic flux through the cap of the bag shown in Fig. 16-11b, and also bounded by the top rim? (*c*) What is the total magnetic flux through the entire bag with the cap on?

SOLUTION   *a*   Since the magnetic field is perpendicular to the circular bottom of area A, the magnetic flux through it alone is $B_0 A$. The magnetic field lines are everywhere parallel to the curved side, and thus the magnetic flux through this portion of the closed surface is zero. Consequently, the *total* flux through this surface bounded by the top rim is $B_0 A$.

b   When the surface bounded by the circular rim at the top is taken to be the flat cap, as in Fig. 16-11b, the magnetic flux is again $B_0 A$, exactly as found in part (*a*) but for a *different surface* bounded by the *same* circular *loop*.

Actually, the result found here for a simple special example holds in general: the magnetic flux of any field, uniform or nonuniform, through *any* closed loop, whatever its shape and quite apart from whether the loop lies in a single plane or not, is independent of the surface of which the loop is a boundary (see Fig. 16-12).

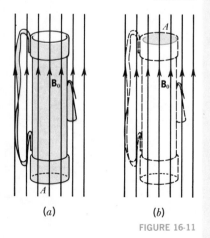

(*a*)   (*b*)

FIGURE 16-11

FIGURE 16-12

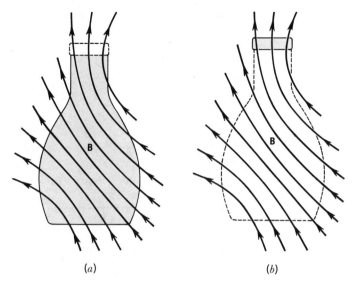

(a)   (b)

c The total magnetic flux through the entire closed bag with its   Magnetic flux through an entire closed
cap on is *zero*. This arises from the fact that as much flux goes   surface is zero.
out through the cap as comes in through the bottom. If we choose
the outward direction as the positive normal of the surface, the
flux out of the top is positive and the flux into the bottom is
negative. Graphically, as Fig. 16-11 shows, this corresponds to
the same number of magnetic field lines leaving the top as entering the bottom of the golf bag. It can be shown that in general
the *net magnetic* flux through any closed surface, whatever its
shape, is zero (see Fig. 16-12).

Faraday's and related experiments[1] show that a *changing flux* through   **16-4 FARADAY'S LAW OF**
a closed loop produces an electromotance around that loop (which   **INDUCED ELECTRIC FIELDS;**
in turn sets up an induced current if the loop happens to be a   **LENZ' LAW**
physical conductor). If the flux is constant, then regardless of how
the magnetic field is changing, there is no electromotance and no   [1]Notably those of Joseph Henry (1797–1878) in the
induced current. But if, for whatever reason, the magnetic flux does   United States. Henry actually observed induced
vary with time, an electromotance is established, described by the   currents before Faraday, but Faraday was the first
simple relation   to publish his results.

$$\langle \mathcal{E} \rangle = \frac{\Delta \Phi_B}{\Delta t}$$ [16-9]

This is Faraday's *law of induction*. It says that a time-varying flux Φ_B
through a closed loop produces around the loop an electromotance

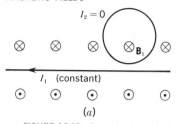

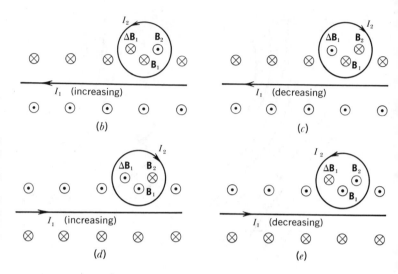

FIGURE 16-13 A circular loop of wire is placed near a long straight wire carrying current I_1 so that the magnetic field B_1 from the current in the wire is everywhere perpendicular to the plane of the loop. The symbol ⊗ indicates a magnetic field directed into the plane of the paper while the symbol ⊙ represents a magnetic field out. The induced current I_2 in the loop produces an induced magnetic field B_2. (a) If the current I_1 is steady, no current is induced in the loop. (b) But if the current I_1 in the direction shown is increasing, the *change* in the magnetic field B_1 within the loop has the direction shown by ΔB_1 and the induced current I_2 has the sense shown. As a result, the magnetic field B_2 has a direction opposite to that of ΔB_1. This as well as the other situations shown in (c) to (e) are described by Lenz' law.

$\mathcal{E}$ whose average value over a time interval Δt is given by the time rate of change of Φ_B.

Faraday's law [16-9], gives the magnitude of $\mathcal{E}$, but it leaves open the question of the direction (sense) of the current. There are just two possibilities; and the rule for deciding between them was arrived at by the physicist H. F. E. Lenz (1804–1865) in Russia, who repeated Faraday's induction experiments.

Lenz' law for the direction of an induced current

An induced electromotance has the sense such that any induced current I_2 produces a new magnetic flux Φ_{B2} opposite to the original *change* of magnetic flux $\Delta\Phi_{B1}$.

This rule, known as Lenz' law, is tested as shown in Fig. 16-13. Here current is induced in a loop of wire by the field of a long straight wire in the same plane. Using the right-hand rule for magnetic fields (Fig. 15-11), the reader should verify that in every case the observed current I_2 sets up a magnetic field $\mathbf{B}_2$ which is opposite in direction to the *change* of the inducing field $\Delta\mathbf{B}_1$. Since the circular area remains fixed, this means that Φ_{B2} and $\Delta\Phi_{B1}$ have opposite signs.

Lenz' law merely an application of energy conservation

Despite its tricky look, Lenz' law is nothing more than the statement that the induction process does not violate conservation of energy. Recall first that we can associate energy with an electric field, as shown in Section 13-7. Similarly, one can associate energy with a magnetic field. We know that when a current is induced in a conductor by a changing magnetic flux, energy appears in the heating of the conductor. The energy must come from somewhere. Now, if we assume the magnetic field produced by the induced current to be in the *same* direction as the change in the magnetic field which produces it, in contradiction to Lenz' law, then the *net* magnetic field would show an enhancement over its magnitude in the absence of an induced current. Therefore, the energy stored in the magnetic field would also be enhanced. But, to account for the energy appearing as heating the conductor, the magnetic field energy must show a diminution, not an enhancement. Thus our assumption must be wrong, and Lenz' law must be correct.

EXAMPLE 16-7

A uniform magnetic field **B** is perpendicular to the plane of the circular loop shown in Fig. 16-14. The initial magnitude of **B** is 2.0 T, and its initial direction is out of the paper, along the positive z axis. We shall take this to be the positive side of the loop's surface. The area of the loop is 0.3 m².

On the first set of axes is plotted the variation of B_z with time. Use the second set to plot Φ_B, and the third set to plot $\mathcal{E}$, as functions

FIGURE 16-14

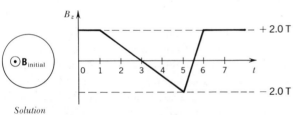

Solution

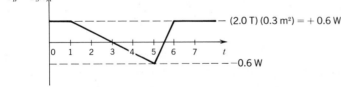

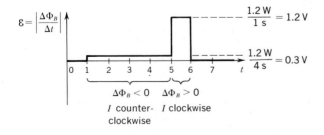

of time. Include numerical values for Φ_B and $\mathcal{E}$ and indicate the range in time for which the induced current is clockwise and counterclockwise.

SOLUTION   See Fig. 16-14.

EXAMPLE 16-8

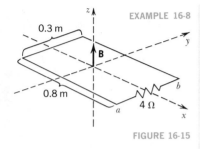

FIGURE 16-15

A 4.0-Ω resistor is connected between points a and b in the rectangular wire loop shown in Fig. 16.15. A uniform magnetic field in the direction of the positive z axis has an initial magnitude 1.5 T. It is then *reduced* uniformly to zero in 0.4 s. (*a*) What is the average electromotance induced in this loop during this 0.4-s interval? (*b*) During this interval is the direction of the current from a to b or from b to a? Explain your reasoning. (*c*) What is the average power dissipated in the Ohmic resistor during this interval?

SOLUTION   *a*   Using Faraday's law [16-9],

$$\langle \mathcal{E} \rangle = \mathcal{E} = \frac{\Delta \Phi_B}{\Delta t} = \frac{\Delta(B_\perp A)}{\Delta t}$$

$$= \frac{(1.5 \text{ T})(0.3 \times 0.8 \text{ m}^2)}{0.4 \text{ s}} = 0.9 \text{ V}$$

Because the magnetic flux is changed uniformly (constant rate), the instantaneous $\mathcal{E}$ is constant and the same as the average $\langle \mathcal{E} \rangle$.

b   From a to b. Here $\Delta B_\perp$ is in the negative z direction. Therefore according to Lenz' law the induced $\mathbf{B}_2$ should be in the direction of the positive z axis. The induced current in the resistor must therefore be from a to b. (See right-hand rule in Fig. 15-11.)

c $\langle P \rangle = \mathcal{E}I = \mathcal{E}\dfrac{\mathcal{E}}{R} = \dfrac{0.81}{4} \approx 0.2 \text{ W}$

EXAMPLE 16-9

Imagine that the rectangular loop of Example 16-8 is bent into the shape shown in Fig. 16-16. What average electromotance is induced in this loop during the 0.4-s interval in which the 1.5-T uniform field reduces to zero?

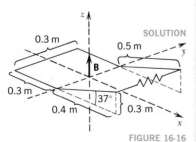

FIGURE 16-16

SOLUTION   Using [16-7], because $B_\perp$ is no longer constant,

$$\langle \mathcal{E} \rangle = \mathcal{E} = \frac{\Delta \Phi_B}{\Delta t} = \frac{\Delta(\Sigma B_\perp \delta A)}{\Delta t}$$

$$= \frac{(1.5)(0.3 \times 0.3) + (1.5)(\cos 37°)(0.3 \times 0.5)}{0.4} = 0.8 \text{ V}$$

which is somewhat smaller than the value found in Example 16-8.

361

+ SECTION 16-5
INDUCED
ELECTROMOTANCE
IN MOVING
CONDUCTORS

FIGURE 16-17

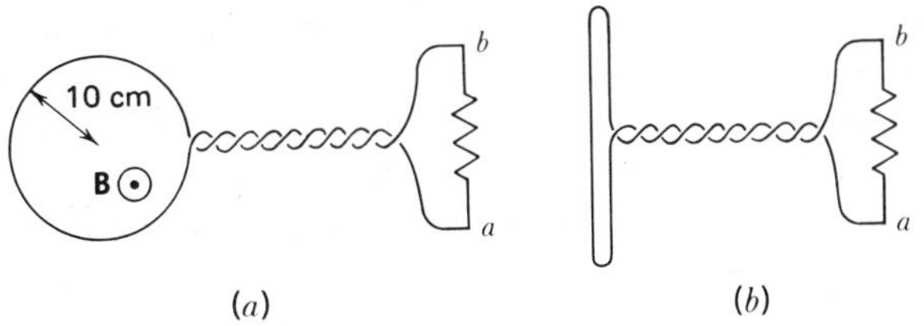

(a)　　　　　　　　(b)

EXAMPLE 16-10

A flexible wire loop initially has the shape of a circle 10 cm in radius. The loop is in the plane of the paper, and a uniform 1.2-T magnetic field points out of the paper (see Fig. 16-17a). Then the loop is pulled at two opposite points and made to collapse in 0.2 s into the shape shown in Fig. 16-17b. (a) What average electromotance is induced across the two terminals a and b during the time the loop is collapsing? (b) During this time what is the direction of the current in the resistance? (c) What average electromotance would have been induced if the magnetic field had been in the plane of the paper?

SOLUTION

a　$\langle \mathcal{E} \rangle = \dfrac{\Delta \Phi_B}{\Delta t} = \dfrac{\Delta(B_\perp A)}{\Delta t} = \dfrac{(1.2)(\pi)(0.1^2)}{0.2} = 0.2 \text{ V}$

b　Initially the flux is out of the paper. As the loop collapses, this flux is reduced to zero. According to Lenz' law, an induced current and induced flux oppose this change; i.e., the induced flux is out of the paper. According to the right-hand rule for magnetic fields (Fig. 15-11), the current in the loop must be counterclockwise. Therefore the current in the resistance must have a direction from a to b.

c　None, because there would be no component of the magnetic field perpendicular to the plane of the loop, no flux, and hence no change of flux.

Faraday's law and Lenz' law are the most general basis for predicting the magnitude and sense of an induced electromotance. Finding the magnitude is simply a matter of computing the change in the magnetic flux, $\Delta(B_\perp A) = \Delta(B \cos \theta\, A)$, whether the flux change arises from change in the magnitude of the magnetic field B through a fixed loop, or through a change in the area A of the loop, or through a change in the relative orientation θ of the magnetic field and the loop, or any combination of the above. Faraday's and Lenz' laws are new and distinctive electromagnetic effects which cannot be derived in their general form from the electric and magnetic effects

+ **16-5　INDUCED ELECTROMOTANCE IN MOVING CONDUCTORS**

Faraday's law arrived at by considering the
electric and magnetic forces on charged
particles in a moving conductor

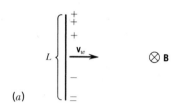

(a)

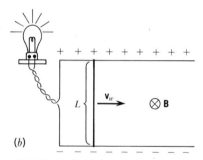

(b)

FIGURE 16-18 (a) When a piece of straight
metal wire is moved with constant speed
in a direction perpendicular to a uniform
magnetic field **B**, the magnetic force on the
charged particles in the wire produces an
excess of the free electrons at one end of
the wire and a deficiency at the other.
Equilibrium is reached when the electric
force from these separated charges balances
the magnetic force. (b) When a light bulb
is connected to the two ends of the moving
wire in (a), as shown, the bulb lights, thus
indicating that the moving wire is a source
of energy, or electromotance.

we have discussed previously. Nevertheless, one special example of induced electromotance, which arises when a conductor is moved through a magnetic field, can be analyzed to reach a result exactly equivalent to that given by Faraday's law merely by examining the electric and magnetic forces acting on the charged particles in a moving conductor.

A piece of thin metal wire of length L is moved parallel to itself at constant velocity $\mathbf{v}_w$ (the subscript w denotes wire). The motion takes place in a uniform magnetic field **B**, which is perpendicular both to $\mathbf{v}_w$ and the wire (see Fig. 16-18). We wish to investigate the forces on charged particles in the moving conductor.

Before the wire begins to move, the charged particles in it are all at rest. Then they acquire a component of velocity $\mathbf{v}_w$ (to the right in Fig. 16-18a) and consequently experience a force from the magnetic field. We know from the right-hand rule for magnetic forces (Fig. 15-16) that this horizontal velocity component produces an upward component of magnetic force on the positive ions of the wire and a downward component of force on the electrons. The ions are locked into the wire by internal forces, but some of the electrons are free to move and start drifting down the wire.

As more and more electrons reach the lower end of the wire, they leave more and more unbalanced positive charge at the upper end. Between these two groups of separated charges a Coulomb force arises, which opposes the magnetic force. Eventually the two forces just balance, and the electron drift ceases (compare the similar equilibrium achieved in the Hall effect, Section 15-8). In the final steady state all charged particles are moving horizontally to the right at constant speed v_w. They experience a vertical magnetic force of magnitude qv_wB balanced by a vertical electric force of magnitude qE in the opposite direction.

No electromotance appears across the moving piece of wire. The sole forces acting are either workless (the magnetic force) or conservative (the Coulomb electric force), and electromotance is work done by a nonconservative force. However, there is a drastic change when this same wire is made to slide between the edges of two parallel conducting plates connected in turn through a light bulb (Fig. 16-18b). The bulb lights!

The electromotance now found around the expanding loop owes its existence to there now being a closed conducting circuit and work done by some nonconservative force. If, for example, the bulb were taken out, the piled-up charges at the end of the traveling wire would

363

+ SECTION 16·5
INDUCED
ELECTROMOTANCE
IN MOVING
CONDUCTORS

merely spread themselves out over the two plates until a new balance between the Coulomb and magnetic forces was attained. The situation would then be the same as previously. But with the light bulb and its resistive load between the plates, we have a segment, the left end of the circuit, through which electrons can be accelerated by the Coulomb field *without* being opposed by the magnetic field. The kinetic energy they gain is turned into random thermal energy, which lights the bulb.

As the electrons move through the bulb, the electric field in the entire region between the plates is diminished. This makes the magnetic force on electrons in the moving wire greater than the Coulomb electric force, so that additional electrons are deflected onto the lower plate. An equilibrium ensues in which electrons circulate steadily around the expanding loop.

But what nonconservative force does the work which accounts for the energy dissipated in the bulb?

To keep the wire moving to the right at the constant speed v_w, it is now necessary to apply a constant external force. We see this from Fig. 16-19. The electrons traveling through the moving wire now have two velocity components: the component $\mathbf{v}_w$ to the right and the component $\mathbf{v}_m$ vertically down, which represents the magnetic drift. The resultant velocity $\mathbf{v}$ makes an angle θ with the horizontal; therefore the magnetic force $\mathbf{F}_m$ makes an angle θ, with the vertical and thus has an unbalanced component to the left. An external force to the right is required to cancel that component in order to keep the velocity $\mathbf{v}$ constant.

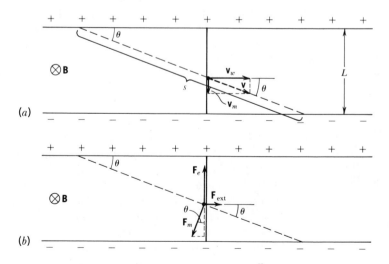

(a)

(b)

FIGURE 16-19 (*a*) The path of an electron drifting from the edge of the upper conducting plate, through the *moving* wire, to the edge of the lower plate is the dashed straight line shown. The velocity $\mathbf{v}$ of the drifting electrons has two components, $\mathbf{v}_w$ perpendicular to the moving wire and arising directly from the motion of the wire and $\mathbf{v}_m$ parallel to the wire and arising from the magnetic effects. (*b*) The free electrons in the wire drift with constant velocity $\mathbf{v}$: the magnetic force $\mathbf{F}_m$ (perpendicular to both $\mathbf{v}$ and $\mathbf{B}$) is balanced by the combined electric force $\mathbf{F}_e$ from the separated charges on the parallel plates and the external force $\mathbf{F}_{ext}$ applied to keep the wire moving at constant velocity.

It is this external force $\mathbf{F}_{ext}$ which does work and is the source of the loop's electromotance. Although, from the macroscopic viewpoint, $\mathbf{F}_{ext}$ is mechanical, the electrons feel it by virtue of their electric charges, and so it may be classified as an electric force. In addition it is a nonconservative force, doing a net amount of work over each electronic round trip.

External force is responsible for the electromotance.

To calculate the electromotance $\mathcal{E}$ arising from the force $\mathbf{F}_{ext}$, we note that an electron travels from the upper to the lower plate along a diagonal path of length $s = L/(\sin\theta)$ (refer to Fig. 16-19). The constant component of $\mathbf{F}_{ext}$ along this path is $F_{ext}\cos\theta = F_m\sin\theta\cos\theta = qvB\sin\theta\cos\theta$, where we have made use of the fact that $\mathbf{F}_{ext}$ balances the horizontal component of $\mathbf{F}_m$. Hence, the work per unit charge done by the external force is

[16-10]
$$\mathcal{E} = \frac{W}{q} = \frac{(qvB\sin\theta\cos\theta)[L/(\sin\theta)]}{q} = vBL\cos\theta = v_w LB$$

But $v_w L$ is the area swept out per second by the traveling wire; i.e., it is the rate of change $\Delta A/\Delta t$ of the area of the conducting loop. Then, remembering that B is constant, we have

$$\mathcal{E} = B\frac{\Delta A}{\Delta t} = \frac{\Delta(BA)}{\Delta t} = \frac{\Delta\Phi_B}{\Delta t}$$

This is Faraday's law, reached from considerations of force and of energy conservation. Needless to say, we have in no sense derived the law in its universal form but only retrieved it in a special case.

Finally, this same analysis also predicts Lenz' law. Let us choose the normal to the plane of the loop to be in the direction of **B**. Then as the vertical wire moves to the right, $\Delta\Phi_B = B\,\Delta A$ increases; but the induced *conventional* current in the loop is counterclockwise, giving rise to a new magnetic field $\mathbf{B}_2$ pointing out of the paper and consequently to a new flux Φ_{B2} opposite to the original *change* in flux, just as described by Lenz' law.

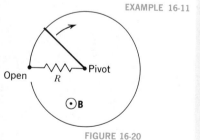

FIGURE 16-20

EXAMPLE 16-11 The electric circuit in Fig. 16-20 consists of three parts: a slender conducting rod 0.4 m in length pivoted at one end, a circular conducting track along which the other end of the slender rod is free to slide, and a resistor R connected from the central pivot to the track. The circular track is broken at a point on one side of the resistor connection, as shown in the figure, and the rod initially rests on the track just on the other side of the resistor connection. A uniform magnetic field of magnitude 0.5 T is perpendicular to the plane of the circuit (out of the paper in Fig. 16-20).

365

+ SECTION 16-5
INDUCED
ELECTROMOTANCE
IN MOVING
CONDUCTORS

If the rod is rotated at the uniform rate of 0.2 rev/s in the clockwise sense, (a) what is the magnitude of the electromotance induced in the closed portion of the circuit, and (b) what is the direction of the induced current in the resistor?

SOLUTION

a $\quad \langle \mathcal{E} \rangle = \mathcal{E} = \dfrac{\Delta \Phi_B}{\Delta t} = B_\perp \dfrac{\Delta A}{\Delta t} = B_\perp \dfrac{\pi r^2}{T}$

$$= \dfrac{(0.5)(\pi)(0.4^2)}{5} \approx 0.05 \text{ V}$$

b   As the rod rotates, the magnetic flux in the closed circuit increases out of the paper. Therefore the induced current, as prescribed by Lenz' law, creates a magnetic field into the paper so as to counteract the above increase in flux. To create this new field the induced current must flow from the pivot through the resistor to the outer circular track.

At time $t = 0$ a rectangular loop of wire, 0.2 by 0.6 m, is in the xy plane in the position I shown in Fig. 16-21. The uniform magnetic field in this region points in the direction of the $+z$ axis and has the magnitude $B_0 = 0.3$ T. The loop is then rotated uniformly in 0.04 s from position I to position II shown in Fig. 16-21. One 0.6-m side of the loop, held fixed, is the axis of rotation. (a) What is the average electromotance during this 90° rotation? (b) What is the maximum value of the electromotance during this 90° rotation? (c) Write a general algebraic expression for the electromotance in the loop at some arbitrary time t after the loop starts to rotate ($0 < t < 0.04$ s).

EXAMPLE 16-12

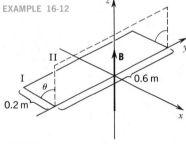

FIGURE 16-21

a   During this 90° rotation $B_\perp$ changes by the full value B_0. Thus

SOLUTION

$$\langle \mathcal{E} \rangle = \dfrac{\Delta \Phi_B}{\Delta t} = \dfrac{(\Delta B_\perp)A}{\Delta t}$$

$$= \dfrac{(0.3)(0.2 \times 0.6)}{0.04}$$

$$= 0.9 \text{ V}$$

b   The maximum value of $\mathcal{E}$ will occur when the outer 0.6 m wire is moving perpendicular to the magnetic field, i.e., at $t = 0.04$ s, when the loop lies in the yz plane. We can see this most easily by noting that only at this instant when the outer wire moves at right angles to the magnetic field do the conduction electrons experience a maximum magnetic force. (Conversely, at the start, when the loop is in the xy plane and the outer wire is moving parallel to the magnetic field, the conduction electrons experience no magnetic force.) Therefore, using [16-10],

$$\mathcal{E}_{max} = vB_0a + 0 + 0 + 0$$

where the first term represents the contribution from the moving 0.6-m wire. There is zero contribution from the other 0.6-m wire (it is at rest) and zero contribution from the two 0.2-m wires (in these wires the magnetic force $\mathbf{F}_m$ is ineffective because it never has a component parallel to the motion of the wire). Expressing the speed v in terms of the time Δt it takes the wire to move one-quarter of the way (90°) around a circle of radius $b = 0.2$ m, we have

$$\mathcal{E}_{max} = \frac{1}{4}\frac{2\pi b}{\Delta t}B_0a$$

$$= \frac{1}{4}\frac{(2\pi)0.2}{0.04}(0.3)(0.6) = 1.4 \text{ V}$$

c The component of the wire's velocity *perpendicular to the magnetic field* is $v_w = v\sin\theta$, where θ, initially zero, increases uniformly with time. If T is the time (period) for one complete revolution (360°), then $\theta/t = 2\pi/T$, or $\theta = (2\pi/T)t = \omega t$. Here ω, the angular velocity, is the number which multiplies the time to give the angular displacement. Therefore, for any time t we have[1]

$$\mathcal{E} = v_wBL = (v\sin\theta)B_0a = B_0av\sin\omega t$$

[1] Using calculus, this result can also be derived directly from Faraday's law. The magnetic flux varies with time as $\Phi_B = B_\perp A = (B_s\cos\theta)ab = B_0ab\cos\omega t$, where $\omega = 2\pi/T = 2\pi/(2\pi b/v) = v/b$. Therefore, $\mathcal{E} = |d\Phi_B/dt| = B_0ab\omega\sin\omega t = B_0av\sin\omega t$

+ 16-6 MAXWELL'S LAW OF INDUCED MAGNETIC FIELDS

Faraday's law: changing magnetic flux produces electric field; Maxwell's law: changing electric flux produces magnetic field

In 1864 James Clerk Maxwell (1831–1879) conjectured that the relationship between electricity and magnetism is in fact reciprocal. Not only does a changing magnetic field create an electric field (as shown in Fig. 16-22a), a changing electric field creates a magnetic field (as shown in Fig. 16-22b). Also on the basis of symmetry he postulated that this induced magnetic field would be related to the

FIGURE 16-22 (a) Faraday showed experimentally that a *change* in magnetic field $\Delta\mathbf{B}_1$ induces an electric field $\mathbf{E}_2$. (b) Maxwell hypothesized that a *change* in electric field $\Delta\mathbf{E}_1$ induces a magnetic field $\mathbf{B}_2$.

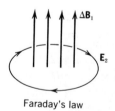

Faraday's law

(a)

Maxwell's hypothesis

(b)

changing electric flux by

$$\frac{\Delta\Phi_E}{\Delta t} = \frac{k_e}{k_m}\Sigma B_\parallel \, \delta s$$ [16-11]

Apart from the constant factor k_e/k_m, the ratio of the fundamental electric and magnetic constants, this is just like Faraday's law

$$\frac{\Delta\Phi_B}{\Delta t} = \Sigma E_\parallel \, \delta s$$

with the roles of **E** and **B** reversed. Maxwell's brilliant guess about this reciprocal nature of electric and magnetic fields was confirmed some 20 years later when Hertz produced electromagnetic waves in the laboratory (Chapter 20).

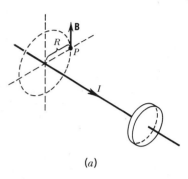

(a)

We can get some idea of how Maxwell was led to [16-11] by considering the charging of a capacitor. Suppose a steady current I exists in the wire of Fig. 16-23a, thus charging the capacitor. It is reasonable to suppose that at outside points which are *far up or down the wire* from the capacitor the current produces the same magnetic field as would exist if the wire were not interrupted. That is, at such points, we have [15-9]

$$B = k_m\frac{2I}{R}$$

with the direction of the field being given by the right-hand rule for magnetic fields. Of course, [15-9] for B cannot be applied near the capacitor since no physical current flows between the plates.

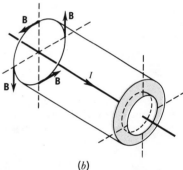

(b)

FIGURE 16-23 (a) When a parallel-plate capacitor is charged by a current in a long straight wire, the magnetic field **B** far from the break in the wire is given by [15-9]. (b) The golf-bag surface used to relate the magnetic field **B** far from the capacitor to the change in electric field ΔE within the capacitor.

There is, however, a relationship between the current in the wire and the rate of change of the electric field **E** in the space *between the plates*. Suppose the charge magnitude on either plate is built up from 0 to ΔQ in time Δt. Then, according to [13-7], the electric field between the plates goes from 0 to $(4\pi k_e \Delta Q)/a$, a being the area of either plate. Hence,

$$\frac{\Delta E}{\Delta t} = \frac{4\pi k_e \, \Delta Q/a}{\Delta t} = 4\pi k_e I/a$$

or

$$\frac{\Delta E}{\Delta t}a = 4\pi k_e I$$ [16-12]

The left side of this equation is simply the time rate of change of the *electric* flux, $\Delta\Phi_E/\Delta t$, through a circular surface of unchanging area a lying between and parallel to the plates. Here we define electric flux, $\Phi_E = \Sigma E_\perp \, \delta A$, in the same fashion as we did magnetic flux, $\Phi_B = \Sigma B_\perp \, \delta A$. Indeed, the left side of [16-12] represents the entire electric flux through the golf-bag surface of Fig. 16-23b, since the electric field vanishes except between the capacitor plates. Now the circle which constitutes the boundary of this surface (the top rim of the bag) is far from the capacitor, and [15-9] for B holds along it. Thus we have

$$\Sigma B_\parallel \, \delta s = 2\pi R \left(\frac{k_m 2I}{R} \right) = 4\pi k_m I$$

around the bag rim. Substituting this result in the right side of [16-12], we obtain

$$\frac{\Delta \Phi_E}{\Delta t} = \frac{k_e}{k_m} \Sigma B_\parallel \, \delta s$$

over bag   around bag
surface   rim

We have arrived at Maxwell's induction law, [16-11] for a *particular* physical situation (current in a long straight wire charging a capacitor) and for a *special* surface (a long golf bag). However, as Maxwell hypothesized and as experiments overwhelmingly corroborate, [16-11] has the same generality as Faraday's law: it applies to *all* physical situations and *all* surfaces. Initially, this was strictly a conjecture by Maxwell. For example, even in the above derivation it is not obvious that [16-11] should apply to the bottom surface of the golf bag (the circular piece between the plates of the capacitor; [15-9] for B used in the above derivation certainly does not hold so near the break in the wire. The fact that [16-11] nevertheless does apply here means that the *changing electric flux acts like the missing current I between the plates of the capacitor.*

Changing electric flux equivalent to an ordinary current

Regardless of whether a real current I is present or not, a change of electric flux through an arbitrary surface will induce a magnetic field. For the direction of the induced magnetic field, one has the relationship pictured in Fig. 16-24. The magnetic field in the vicinity of the capacitor is the same, both in magnitude and direction, as the magnetic field at all other points along the wire. Note that the relationship between the direction of $\Delta \mathbf{E}$ and the sense of $\mathbf{B}$ shown in Fig. 16-24 is the same as in Fig. 16-23b. This relationship is therefore described by the right-hand rule for magnetic fields, where the direction of the change in electric field $\Delta \mathbf{E}$ replaces the direction of the current I.

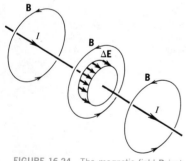

FIGURE 16-24 The magnetic field **B** just outside the parallel-plate capacitor of Fig. 16-23 is identical to that found far from the capacitor. The changing electric field between the capacitor plates produces a magnetic field, just as the current in the wire away from the capacitor produces such a field.

EXAMPLE 16-13

Using Maxwell's law, show that the magnetic field between the parallel plates of Fig. 16-23 and tangential to circles concentric with the plates is (a) zero at the center of the capacitor, (b) directly proportional to the radial distance R for $R < R_0$, but (c) given by [15-9] at $R = R_0$, where R_0 is the radius of the capacitor plates.

Consider a circle of radius R within the capacitor concentric with the capacitor plates. For this curve

$$\frac{\Delta \Phi_E}{\Delta t} = \frac{k_e}{k_m} \Sigma B_{\parallel} \, \delta s$$

$$\frac{\Delta E \pi R^2}{\Delta t} = \frac{k_e}{k_m} B 2 \pi R$$

Using [13-7] for E, and assuming the charge per unit area changes by $\Delta \sigma$ in time Δt,

$$\frac{4 \pi k_e \, \Delta \sigma \pi R^2}{\Delta t} = \frac{k_e}{k_m} B 2 \pi R$$

or

$$B = k_m \frac{2I}{R_0} \frac{R}{R_0} \qquad \text{where } I = \pi R_0^2 \frac{\Delta \sigma}{\Delta t}$$

a   For $R = 0$, $B = 0$.
b   For $0 < R < R_0$, $B \propto R$.
c   For $R = R_0$, $B = k_m 2I/R_0$.

Faraday's law of induced electric fields describes the electric field produced by a changing magnetic field

$$\langle \mathcal{E} \rangle = \frac{\Delta \Phi_B}{\Delta t} \tag{16-9}$$

where the electromotance $\mathcal{E}$ is given by

$$\mathcal{E} = \Sigma E_{\parallel} \, \delta s \tag{16-1b}$$

and the magnetic flux Φ_B is given by

$$\Phi_B = \Sigma B_{\perp} \, \delta A \tag{16-7}$$

According to Lenz' law, an induced electromotance has the sense such that any induced current I_2 produces a new magnetic flux Φ_{B2} opposite to the original *change* of magnetic flux $\Delta \Phi_{B1}$.

Maxwell's law of induced magnetic fields [16-11] describes the magnetic field produced by a changing electric field. The sense of the induced magnetic field is given by the usual right-hand rule for magnetic fields, with the direction of the current I replaced by the direction of the change in electric field $\Delta\mathbf{E}$.

PROBLEMS

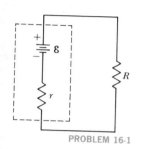

PROBLEM 16-1

16-1 Thermal energy is generated in the 0.10-Ω resistor R shown in the circuit at the rate of 10 W by connecting it to a battery having an electromotance of 1.5 V. (*a*) What is the current in the circuit? (*b*) What is the battery's internal resistance r? (*c*) What electric potential difference exists across the 0.10-Ω resistor? (*d*) How much power is being supplied by the battery? (*e*) Show that the power supplied by the battery is consistent with energy conservation in that it is equal to the rate at which thermal energy is generated in the entire circuit.

16-2 A 2-A current exists in the 3-Ω resistor shown in the circuit. The battery has an internal resistance of 1 Ω. (*a*) How much power is dissipated in the 3-Ω resistor? (*b*) How much power is dissipated in *each* 8-Ω resistor? (*c*) Use the principle of energy (or power) conservation to determine the electromotance $\mathcal{E}$ of the battery. (*d*) What is the electric potential difference between the terminals A and B of the battery?

16-3 (*a*) What is the current in the battery shown in the circuit? (Take the internal resistance of the battery to be negligible.) (*b*) If the electric potential at point A is zero (ground), what is the electric potential at point B? (*c*) At point C? (*d*) At point D? (*e*) How large a resistance would have to be connected from A to C in order to increase the battery current to 4 A?

16-4 A galvanometer, a current-measuring device, is connected to a 70-Ω resistor as shown in the figure. The resistor has a variable tap, or connection. The internal resistance of the galvanometer is 90 Ω. What should be the resistance between points B and C if only 25 percent of the current drawn from the battery is to go through the galvanometer?

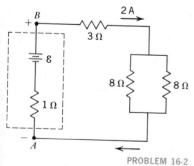

PROBLEM 16-2

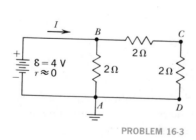

PROBLEM 16-3

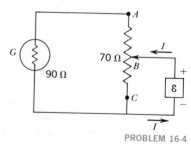

PROBLEM 16-4

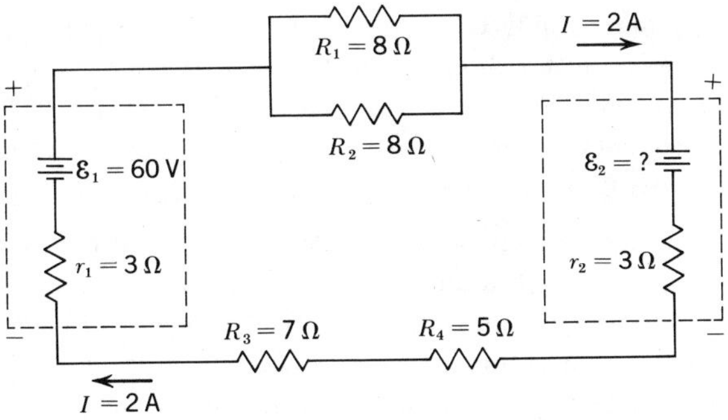

PROBLEM 16-5

(*a*) Use the conservation-of-power principle to compute the value of the unknown electromotance $\mathcal{E}_2$ in the circuit shown. Both sources of electromotance have an internal resistance of 3 Ω. (*b*) What is the electric potential difference across the terminals of the source $\mathcal{E}_2$? (*c*) Across the terminals of $\mathcal{E}_1$? 16-5

The magnitude and direction of a nonconservative electric field E' for various points along the line from A to E are shown in the figure. Calculate the electromotance between points (*a*) A and B, (*b*) A and C, (*c*) C and D, (*d*) D and E, and (*e*) A and E. 16-6

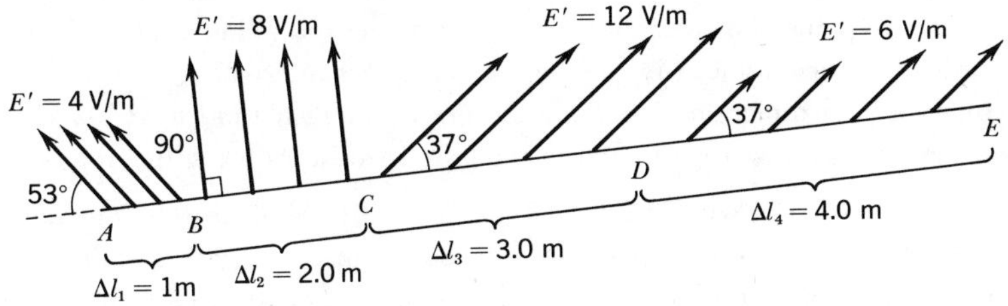

PROBLEM 16-6

A surface has the form of a *roof* and end *walls* above an attic floor. The area of the surface is 100 cm^2 for the inclined roof (shown in the figure) and 24 cm^2 for the end walls, for a total of 124 cm^2. What is the magnitude of the total flux through this surface if it is held in a constant northward magnetic field of 2.0 T so that plane of the attic floor is *vertical* and contains (*a*) only east-west lines, (*b*) only north-south lines? (*c*) If the surface is held so that the plane of the floor is *horizontal* and the vertical walls contain only east-west lines? (The floor is *not* considered part of the surface.) 16-7

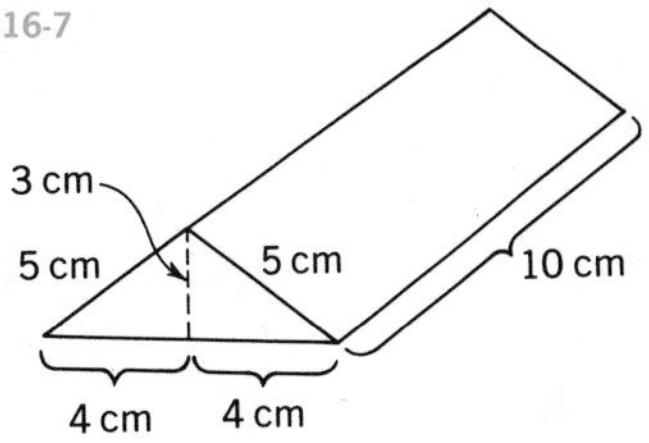

PROBLEM 16-7

A uniform magnetic field everywhere has the direction of the $+z$ axis, and its magnitude B varies with time as shown in the figure. Consider a wire loop of area 0.7 m^2 in the xy plane, $z = 0$. What is the average elec- 16-8

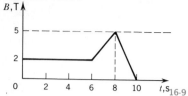

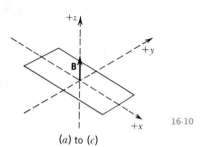

(a) to (c)

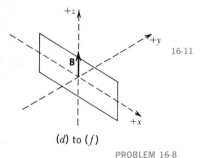

(d) to (f)

PROBLEM 16-8

tromotance induced in this loop (mangitude only) during the interval (a) $t = 0$ to 6 s, (b) $t = 6$ to 8 s, and (c) $t = 8$ to 10 s? Now consider a rectangular loop of wire in the xz plane, $y = 0$. What is the average electomotance induced in the loop (magnitude only) during the interval (d) $t = 0$ to $t = 6$ s, (e) $t = 6$ to $t = 8$ s, (f) $t = 8$ to $t = 10$ s?

16-9

Consider the rectangular loop shown in Fig. 16-10. Take its dimensions to be 0.10 by 0.30 m and assume the uniform magnetic field along the $+z$ axis to have the magnitude 0.50 T. The loop is turned about an axis lying along the y axis, and the loop goes from the orientation shown in Fig. 16-10a to that in Fig. 16-10b, with $\theta = 30°$, in a time interval of $\frac{1}{30}$ s. Another $\frac{2}{30}$ s is required until it assumes the orientation shown in Fig. 16-10c. What are the magnitude and sense of the average induced electromotance (a) over the first $\frac{1}{30}$ s, (b) over the next $\frac{2}{30}$ s, and (c) over the entire $\frac{3}{30}$-s interval?

16-10

A uniform magnetic field **B** is perpendicular to the plane of the circular loop of wire shown in the figure. The time variation of B_z is shown in the graph. The loop's area is 0.5 m². Plot the electromotance induced in the loop as a function of time, indicating a clockwise $\mathcal{E}$ as positive and a counterclockwise $\mathcal{E}$ as negative.

16-11

A 5.0-Ω resistor is connected between points a and b in the rectangular loop of wire shown in part (a) of the figure. The loop is 0.06 by 0.02 m. A uniform magnetic field of magnitude 3.0 T points in the direction of the positive z axis. One-half of the loop is then bent in 0.03 s into the position in the xz plane shown in part (b) of the figure. (a) What is the average electromotance induced in the loop in this 0.03-s interval? (b) During this interval is the direction of the induced current from a to b or b to a? (c) What is the average power dissipated in the Ohmic resistor during this interval?

16-12

A circular loop of wire of area 0.5 m² lies in the xy plane. The magnetic field **B** in the region of space initially points in the negative z direction and has a magnitude 3.5 T. During a time interval of 40 s the z component of **B** changes at a uniform rate from $B_z = -3.5$ T to $B_z = +3.5$ T. What is the magnitude and sense of the average electromotance over (a) the first half of this time interval, (b) the second half, and (c) the full interval?

PROBLEM 16-10

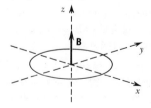

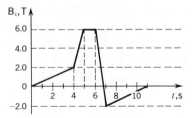

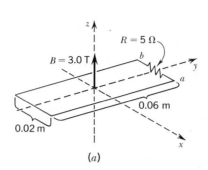

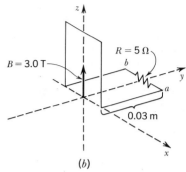

PROBLEM 16-11

Suppose that all the dimensions of the rectangular conducting loop shown in Fig. 16-15 are shrinking uniformly at the rate of 1.0 percent per second. The field **B** along the z direction has the constant magnitude 2.0 T. It is intended that *no* electromotance be induced in the loop. An additional magnetic field, uniform over the dimensions of the loop but increasing steadily with time, is available. (*a*) In what direction must the magnetic field be directed? (*b*) At what rate must the magnetic field change with time in order for the electromotance to be zero? 16-13

A small rectangular coil of 500 turns of wire and area 20 cm² is rotated at 1,800 rpm in a uniform but unknown magnetic field as shown in the figure. The electromotance induced in the coil creates an open-circuit electric potential difference across the terminals of the coil which varies from 0 to a maximum value of 0.02 V. What is the magnitude of **B?** 16-14

A jet plane has a perfectly conducting wire extending from the tip of its left wing to the tip of its right wing. (*a*) Is an electromotance induced in the wire as the plane flies through the Earth's magnetic field? (*b*) Can this effect be used as a source of power for electric lights, even a tiny flashlight bulb, on the jet? Explain your reasoning. 16-15

A changing magnetic flux through a uniform circular conducting ring induces an electromotance of 24 V. What is the electric potential difference between two points on the ring at opposite ends of a diameter? 16-16

A thin conducting rod 1.0 m in length slides over two parallel conducting rods connected to a lamp bulb having a resistance of 10 Ω, as shown in Fig. 16-18*b*. The rod's initial speed to the right is 2.0 m/s, and the uniform magnetic field into the plane of paper has a constant magnitude of 0.50 T. What is (*a*) the rate at which the magnetic flux through the conducting loop initially changes with time, (*b*) the magnitude of the electromotance induced in the loop at this instant, (*c*) the magnitude of the current induced in the loop at this instant, (*d*) the sense (clockwise or counterclockwise) 16-17

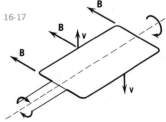

PROBLEM 16-14

of the induced current, (e) the rate at which energy is dissipated in the resistance of the lamp at this instant, and (f) the magnitude and (g) direction of the force acting on the rod at this instant?

16-18   The uniform electric field between the plates of a parallel-plate capacitor initially has the magnitude of 2.0×10^4 V/m. The electric field then drops to zero in a time interval of $\frac{1}{10}$ s. The radius of the circular capacitor plates is 0.050 m. (a) What is the magnitude of the initial electric flux through a surface lying between, and parallel to, the two capitor plates and having a radius of 0.050 m? (b) What is the average time rate of change of the electric flux through this surface? (c) What is the magnitude of the magnetic field induced at a point between the plates and 0.050 m from the center? (d) What is the sense (clockwise or counterclockwise) of the induced magnetic field if the decreasing electric field has the direction into the surface about whose boundary the magnetic field is induced?

LIGHT AND THE CLASSICAL PARTICLE THEORY

What is light? The next four chapters are devoted to this question. The answer, historically speaking, has barely begun. Once again our starting point must be experiment and observation. Light exhibits a large variety of effects, and the present chapter will explore some of them. Our first idea is to see to what extent we can account for them by a simple and plausible assumption—that light consists of particles.

17-1 SOME FAMILIAR
OBSERVATIONS

Some well-known properties of light

Light represents the *transport of energy,* as anyone who steps from the shadow into sunlight experiences directly. A source of light loses energy, which somehow travels to distant objects, heating them.

Until it encounters an obstacle, *light travels in straight lines.* We have all seen the sharp beam of a searchlight or of light breaking through

FIGURE 17-1 A photograph taken from the moon by Surveyor 7. The two white specks in the dark half of the crescent Earth are the cross sections of two very sharp light beams (from two lasers 500 miles apart). (*Spectra-Physics, Inc.*)

an opening in clouds into a hazy atmosphere (see Fig. 17-1). The term *ray* is commonly used to denote the path of the light from the source to some distant point. That light travels through a single medium, such as air, in perfectly straight lines is taken as obvious when we test the straightness of a ruler by sighting along its edge.

Light is *reflected in simple fashion* when it falls upon a plane smooth surface. Looking at an object in a mirror is essentially like looking at it directly. In fact, experiments with a single ray impinging on a reflecting surface readily establish the following law:

1   The incident ray, the reflected ray, and the normal to the surface at the point where the ray strikes all lie in a single plane, as shown in Fig. 17-2.
2   The angle between the incident ray and the normal is the same as the angle between the reflected ray and the normal (the angle of incidence equals the angle of reflection). Representing these angles by θ and θ', respectively, we write

$$\theta = \theta' \qquad [17\text{-}1]$$

Figure 17-3 shows how an extended object forms an image in a mirror. Here the two points A and B, for example, a man's nose and his raised right hand, are imagined to be light sources which shine on a mirror. Rays from these sources can reach the left and right eyes, E_l and E_r, of another observer by reflection at the mirror. For every ray the angles of reflection and incidence are equal. The eye is naïve in the sense that it interprets any ray entering it as having come straight from the source to it. Therefore, for the second observer it is as if the rays had come from image points A' and B' whose separation and distances from the mirror are exactly the same as those of the two point sources. There is a difference, however; point A' shows that the image is facing the opposite direction, and point B' shows that the image has a raised *left* hand. Thus, left and right are reversed when the image faces the opposite direction.

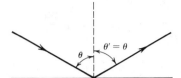

FIGURE 17-2   Law of reflection: the reflected beam lies in the plane of incidence (the plane containing the incident beam and the normal to the reflecting surface), and the angle of incidence θ is equal to the angle of reflection θ'.

Law of reflection

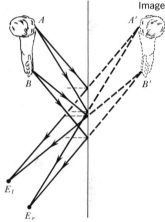

FIGURE 17-3   An observer using his right (E_r) and left (E_l) eyes sees for the image of a man with *right* arm raised a man facing the opposite direction with *left* arm raised. The reflected rays used to determine the image are drawn according to the law of reflection.

EXAMPLE 17-1

A man 6 ft tall stands in front of a mirror hanging on a wall. His eyes are 4 in. below the top of his head. What is the shortest mirror which, without being moved, will permit the man to see his entire body? Where must the mirror be placed? Does it make a difference how far he stands from the mirror?

SOLUTION

For the man to see the top of his head and the tips of his toes, light rays from these points must strike the mirror and be reflected back

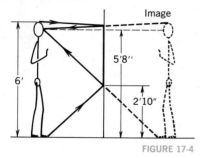

Image

5′8″

6′

2′10″

FIGURE 17-4

to the man's eyes. Since the angle of incidence is equal to the angle of reflection, if we assume the man's forehead, eyes, and toes are all in the same vertical plane, it is easy to see, using Fig. 17-4, that the man will never have to use points on the mirror higher than 5 ft 10 in. (halfway between the height of his eyes and the height of the top of his head) or lower than 2 ft 10 in. (half the height of his eyes). Thus the mirror should be 3 ft tall, and its top should be 5 ft 10 in. above the floor. If the man stands at a different distance from the mirror, the angles of incidence and reflection change, but clearly the required position of the mirror does not.

17-2 THE CLASSICAL PARTICLE THEORY

A classical model for light: light consists of particles.

In this theory, which has existed in one form or another since the time of early Greek civilization, one imagines that light consists of particles emitted by the light source. These particles travel outward through a uniform medium in straight lines and upon striking a reflecting surface behave just like small hard spheres colliding elastically with a massive wall.

There is no difficulty in accounting for energy transport by this model: light energy is the kinetic energy of the particles, which appears as thermal energy when the carriers are captured by an absorbing material. Reflection of light is also easily explained by the assumption of elastic collisions. From Fig. 17-5 we see that the angles θ and θ' are related to the velocity components for incidence and reflection by

$$\tan \theta = v_x/v_y \qquad \text{and} \qquad \tan \theta' = v'_x/v'_y$$

Here the reflecting surface is in the xz plane, the incident ray strikes this surface at the origin, and the incident (unprimed) and reflected (primed) rays are in the xy plane.

The only force which the smooth surface exerts on the light particles is perpendicular to the surface, i.e., along the y direction. Therefore, the velocity components along the x direction are the same

$$v'_x = v_x$$

and the z velocity components are zero both before and after the collision. The y velocity components are in opposite directions, but because the collision is perfectly elastic, they have the same magnitude

$$v'_y = v_y$$

(This is just the reasoning applied in Chapter 10 to the collision of

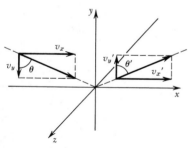

y

v_x

v_y θ

v'_y θ'

v'_x

x

z

FIGURE 17-5 The law of reflection for a particle colliding elastically with a massive (fixed) floor. The surface of the floor is the xz plane, $y = 0$.

an ideal-gas molecule with the container wall.) The last three equations together imply that the angle of reflection is equal to the angle of incidence

$$\theta' = \theta$$

which is precisely the observed result.

17-3 SOME PREDICTIONS OF THE PARTICLE MODEL

If light is indeed a stream of particles, two other effects allied to those just described can be expected. Both are quite independent of the nature and properties of the light "corpuscles."

First, the force exerted on the particles and responsible for their reflection implies an equal and opposite reaction on the illuminated surface. In other words, light should exert a pressure on whatever it strikes. The pressure of light *is* observed and was first detected in delicate experiments by the Russian physicist Peter Lebedev (1866–1912), in 1907. The subtleness of the effect can be gauged from the fact that bright sunlight shining on a mirror of area 1 cm^2 exerts a force of less than 10^{-9} N. This is in contrast with the force of about 10 N on the same area arising from atmospheric pressure, i.e., from molecular bombardment. The extremely minute radiation force is detected with the torsion pendulum, the same highly sensitive instrument with which Cavendish measured the gravitational attraction between objects of laboratory size.

Light exerts a pressure.

How does the force of light on a perfectly reflecting white surface or a mirror compare with the force on a perfectly absorbing black surface?

EXAMPLE 17-2

The question is answered by considering the change in the linear momentum of the incident particles, here imagined for simplicity to strike the surface at right angles. In absorption the particles are brought to rest at the surface, each changing momentum by the amount mv. In reflection the particles are first brought to rest and then accelerated back to their original momentum magnitude, but in the reverse direction, so that the total change in momentum is $2mv$. Since the average force on an object has the same magnitude as the average rate of change of momentum, assuming that the two processes take place in the same time interval, the radiation force for reflection is twice that for absorption. This is confirmed by experiment.

SOLUTION

Query Can you think of a way to design an "interplanetary sailboat" for space travel? Could such a space vehicle accelerate *and* decelerate?

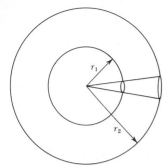

FIGURE 17-6 A point source of light surrounded by two concentric spherical surfaces of radius r_1 and r_2. Because the same number of particles pass through each spherical surface per unit time, the number crossing *per unit area* per unit time must vary inversely as the square of the radius of the spherical surface. At r_2 the number per unit area is smaller than at r_1, but the total area is larger.

Intensity of light from a small source varies inversely as the square of the distance.

The second effect relates to the amount of energy absorbed by a black object at various distances from a small source of light. Certainly, a small light source is more effective in heating an object close to it than one farther away. Using arguments from geometry alone, one can predict from the particle model exactly how heating depends upon the distance between the source and the illuminated object.

Suppose that a point source of light is completely enclosed in an imaginary sphere of radius r_1, as shown in Fig. 17-6. Then all the particles leaving the source radially outward must pass through the spherical surface. These same particles also all pass through a larger concentric sphere of radius r_2. Because the surface area of a sphere $(4\pi r^2)$ varies directly as the square of the radius, it follows that the number of particles per unit time through a unit transverse area, and therefore the thermal energy acquired by a surface, varies inversely as the square of the distance from the point source. This prediction can be checked directly, e.g., by measuring the rise in temperature of a black object over a fixed interval of illumination. The temperature change indicates how much energy is absorbed from the light. Sure enough, the *intensity*, i.e., the amount of energy passing per unit time through a unit area oriented at right angles to the rays, is found to vary precisely as the inverse square of the distance from the point source.

So far the assumption that light sources radiate particles is in agreement with experiment, and our confidence in this simple picture is strengthened. But more questions must be asked and settled by experiment before we can be sure that light really consists of particles.

17-4   THE SPEED OF LIGHT

A basic question is whether a light signal is transmitted instantaneously or whether its speed is finite. The first try at an answer was made by Galileo. He had two observers, A and B, each equipped with a lantern that could be exposed or shuttered. The observers sent signals to each other, first when they were only a short distance apart. Observer A signaled B by exposing his lantern. As soon as B saw the light flash, he signaled back to A by opening the shutter of his lantern. Observer A was to measure the time interval for the round trip. Recognizing that the total elapsed time included the reaction time for the two observers, Galileo repeated the measurements with the two observers separated by several kilometers. Taking the difference in the total time interval for the two separation distances eliminates the effects of reaction time and gives the delay arising from the speed of light.

Galileo failed. As we know, the difference in time interval was only about 10^{-5} s, which is much smaller than the variation in human reaction time from one trial to another. Clearly, to measure the speed of light with any precision, one must either use very large distances, such as astronomical distances, to make the time interval reasonably long, or devise ways of measuring very small time intervals. Both procedures have been applied, and the presently accepted value of the speed of light through a vacuum is

$$c = (2.997925 \pm 0.000001) \times 10^8 \text{ m/s} \approx 3 \times 10^8 \text{ m/s}$$

The measured speed of light,
$c = 3 \times 10^8$ m/s

The speed of light through air is only slightly less.

The measurements show that the speed of light does not depend upon the brightness of the light source. Although this does not necessarily go against the particle hypothesis (or favor it, either) one might imagine that a very hot or bright light source should emit particles of higher kinetic energy than a relatively cool, dim one. But if light does consist of particles, they always travel through empty space at the same speed.

EXAMPLE 17-3

The first measurement of c was made by the Danish astronomer Olaus Roemer (1644–1710) and Christian Huygens, the Dutch scientist. It had been noticed that Io, the innermost moon of the planet Jupiter, showed a rather strange behavior. The period of this moon, as measured by timing its emergence from behind Jupiter, was not constant. Sometimes Io's eclipse ended sooner, sometimes later, than was expected for a regularly orbiting body. The observed periods differed by as much as 30 s out of 42 h. Roemer in 1675 perceived that these variations arose from the finite travel time of light and succeeded in computing the time it takes light to cross the diameter of the Earth's orbit. Then in 1678 Huygens used the first determinations of the length of this diameter to compute the speed of light. How did they do this? (The average diameter of the Earth's orbit about the sun is 3.0×10^{11} m, and the Earth makes about 12 orbits about the sun for every orbit made by Jupiter.)

SOLUTION

In Fig. 17-7 we see the Earth orbiting the sun. Moving in a much larger orbit with a longer period (about 12 Earth years) is the planet Jupiter and its moon with a period of about 42 h. Roemer realized that the time τ between successive appearances of Jupiter's moon is not the same as the period of the moon's orbit. Instead, because the Earth moves, that interval must be corrected for the different distances traveled by the light signals corresponding to the two

An early demonstration that light's speed is finite

FIGURE 17-7

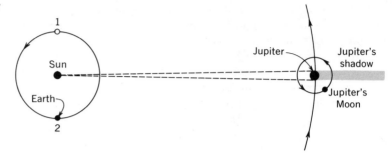

emergences. For position 1, with the Earth moving away from Jupiter,

$$\tau_1 = P + \frac{d_1}{c}$$

where P is true period of the moon, c is the speed of light, and d_1 is the distance the Earth moves in the time τ_1. Then 6 months later, with the Earth at position 2 and moving toward Jupiter's moon,

$$\tau_2 = P - \frac{d_2}{c}$$

The difference in time intervals is

$$\tau_1 - \tau_2 = \frac{d_1 + d_2}{c}$$

or

$$c = \frac{d_1 + d_2}{\tau_1 - \tau_2}$$

The distances d_1 and d_2 can be related to the Earth's orbital speed v_E, or the period T_E ($= 1$ year) and radius R_E of the Earth about the sun,

$$c = \frac{d_1 + d_2}{\tau_1 - \tau_2} = \frac{\tau_1 + \tau_2}{\tau_1 - \tau_2} v_E = \frac{\tau_1 + \tau_2}{\tau_1 - \tau_2} \frac{2\pi R_E}{T_E}$$

Roemer found $\tau_1 + \tau_2$ to be 84 h and $\tau_1 - \tau_2$ to be 30 s. The average radius of the Earth's orbit is now known to be 1.49×10^{11} m. Therefore, the speed of light is

$$c = \frac{84 \text{ h}}{30 \text{ s}} \frac{6.28 \times 1.49 \times 10^{11} \text{ m}}{365 \text{ days} \times 24 \text{ h/day}} = 3.0 \times 10^8 \text{ m/s}$$

Only a crude value for the radius of the Earth's orbit was available to Huygens, who obtained for c the value 2×10^8 m/s. This was, nevertheless, a remarkable accomplishment for 1678.

EXAMPLE 17-4

The first successful terrestrial measurement of c was made by A. H. L. Fizeau (1819–1896) in 1849. The essential part of the apparatus was a toothed wheel which could be spun at high speeds. As Fig. 17-8 shows, with the wheel at rest, light from a source passes between two cogs to travel to a distant mirror and then back along the same path through the opening. When the wheel rotates, the light emerges in bursts. If the time taken by the light bursts to travel a round trip to the mirror and back is the same as the time for the wheel to turn the width of a tooth, the light passes straight through the rotating wheel on return. But at half this speed the light will be stopped, or eclipsed, by the intervening tooth. Fizeau used a wheel having 720 teeth. He found that as the wheel was speeded up, the reflected light was first eclipsed at 12.6 rev/s. The distance from the wheel to the mirror was 8,633 m. What value did he obtain for the speed of light?

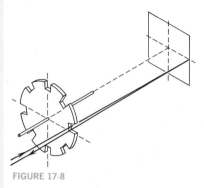

FIGURE 17-8

In 1 s the wheel makes 12.6 rev; therefore, it takes $1/12.6 = 7.94 \times 10^{-2}$ s for one complete revolution. It takes $7.94 \times 10^{-2}/720 = 1.10 \times 10^{-4}$ s for one opening at the beam to be replaced by the adjacent opening, or half this time, 5.50×10^{-5} s, for an opening to be replaced by the adjacent tooth. This is the time for the first eclipse of the reflected light. Since the light traveled a total distance $2 \times 8,633$ m during this time, Fizeau concluded that the speed of the light was

SOLUTION

$$c = \frac{2L}{\Delta t} = \frac{2 \times 8,633 \text{ m}}{5.50 \times 10^{-5} \text{ s}} = 3.15 \times 10^8 \text{ m/s}$$

The Fizeau speed-of-light measurement

(Fizeau's biggest experimental error was in trying to identify the exact rotational frequency corresponding to the eclipses; this accounted for an error in the time Δt.)

17-5 SNELL'S LAW

When light strikes the interface between two transparent materials, such as the surface separating air and water, part is reflected back into the first medium and part is transmitted into the second medium. Unless the light is incident along the normal to the surface, the incident and transmitted rays are generally not in the same

Refraction: bending of a light ray at an interface

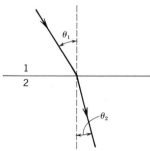

FIGURE 17-9 When a beam of light is incident upon the interface of two media, the light that is transmitted across the interface forms a beam that is generally bent (refracted) toward or away from the normal to the interface. The angle of incidence is θ_1, and the angle of transmission is θ_2.

Snell's law of refraction

[17-2]

Substance	n
Helium	1.000035
Nitrogen	1.0030
Water	1.33
Soda glass	1.48
Heavy flint glass	1.7
Diamond	2.42

TABLE 17-1

FIGURE 17-10 (a) A light beam in air incident upon an air-water interface is refracted toward the normal of the interface upon transmission into the water ($\theta_2 < \theta_1$). (b) A light beam in glass incident upon a glass-water interface is refracted away from the normal to the interface upon transmission into the water ($\theta_2 > \theta_1$).

direction. In this section we set down the basic facts concerning this bending, or *refraction*, of light.

We denote the angle between the incident ray in medium 1 and the line normal to the surface by θ_1. The transmitted ray in medium 2 makes an angle θ_2 with the normal (see Fig. 17-9).

1 It is found by experiment that the incident ray, the transmitted ray, and the normal all lie in a single plane, which also contains the reflected ray. Except for incidence along the normal, the angles θ_1 and θ_2 differ. If medium 1 is air and medium 2 is water, θ_1 exceeds θ_2, as in Fig. 17-10a. On the other hand, if medium 1 is glass and medium 2 is water, θ_2 exceeds θ_1, as in Fig. 17-10b. The transmitted ray is commonly bent toward the normal if the light enters a denser medium and away from the normal if the light goes into a rarer medium.

2 The exact relation between the angles θ_1 and θ_2 was first given in 1621 by Willebrord Snell (1591–1626), a Dutch mathematician. For a given pair of media, the ratio of the sine of the angle of incidence to the sine of the angle of transmission is a constant, independent of both the angle θ_1 and the intensity of the light. Snell's law therefore is

$$\frac{\sin \theta_1}{\sin \theta_2} = n_{21}$$

where n_{21} is a constant characteristic of the two media called the *index of refraction* of medium 2 relative to medium 1. Clearly, from [17-2], if θ_1 exceeds θ_2, as is usually the case when light enters a denser medium, the dimensionless constant n_{21} is greater than 1. Similarly, if medium 2 is optically less dense than medium 1, $\theta_2 > \theta_1$, and $n_{21} < 1$.

If the first or incident medium is a vacuum (approximated closely by air), it is conventional to designate the constant as n_2, or simply

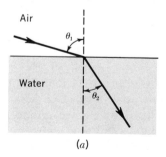

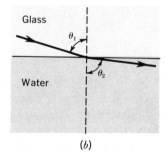

(a)   (b)

as n, and to call this the index of refraction of the second medium. It is then understood that an index with just one subscript, or no subscript, is always relative to *incidence* from a vacuum. Some typical values of n, for standard conditions of temperature and pressure, are listed in Table 17-1.

Refraction of light is reversible in the same way that reflection is. If your friend is visible to you in the mirror, you are also visible to him. Likewise, if a ray goes from A in medium 1 to B in medium 2 via refraction at an interface, a ray goes from B to A by following exactly the same route in the opposite direction (see Fig. 17-11). This implies that it is not necessary to regard θ_1 strictly as the incident angle and θ_2 as the transmitted angle. Because of the reversibility of ray directions, we can equally well make the opposite choice. But if θ_2 is taken as the angle of incidence and θ_1 the angle of transmission, the refractive index used must be n_{12}, the index of refraction of medium 1 relative to medium 2. Since in this case $(\sin \theta_2)/(\sin \theta_1) = n_{12}$, it follows from [17-2] that

$$n_{12} = \frac{1}{n_{21}} \qquad [17\text{-}3]$$

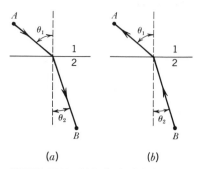

FIGURE 17-11 (*a*) Path of a light beam going from point A in medium 1 to point B in medium 2. Here θ_1 is the angle of incidence, θ_2 is the angle of transmission, and n_{21} is the index of refraction. (*b*) A light beam going from point B in medium 2 to point A in medium 1 also follows the same path but with θ_2 as the angle of incidence and θ_1 as the angle of transmission. Thus from Snell's law the new index of refraction n_{12} is the reciprocal of the previous index n_{21}.

EXAMPLE 17-5

The early Greek philosophers knew the law of reflection, but in their first attempts to write a simple law of refraction they supposed that the angle of refraction θ_2 is directly proportional to the angle of incidence θ_1

$$\theta_2 = n_{12}\theta_1$$

Although this is not the correct formulation, can you show that it is not inconsistent with Snell's law, at least for small angles of incidence and refraction?

SOLUTION

For a small angle θ in Fig. 17-12*a*, $\sin \theta = y/R$ is very nearly the same as the angle in radian measure, $\theta = s/R$. To three significant figures there is no difference between θ and $\sin \theta$ until θ is larger than $8°$.

FIGURE 17-12

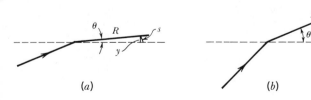

(*a*)   (*b*)

EXAMPLE 17-6   A transparent plate (medium 2) with parallel surfaces separates medium 1 from medium 3, as shown in Fig. 17-13. A ray of light is incident upon the surface between 1 and 2 at the angle θ_1. Find the expression giving the angle θ_3 at which the ray emerges into the third medium in terms of the relative indices of refraction n_{21} and n_{32}.

SOLUTION   For the first refraction,

$$\frac{\sin \theta_1}{\sin \theta_2} = n_{21}$$

and for the second refraction,

$$\frac{\sin \theta_2}{\sin \theta_3} = n_{32}$$

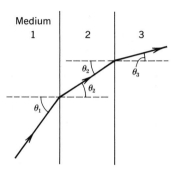

Medium
1   2   3

FIGURE 17-13

Therefore, eliminating $\sin \theta_2$ from these two equations gives

$$\sin \theta_3 = \frac{1}{n_{32}} \frac{1}{n_{21}} \sin \theta_1$$

Now suppose that the first and third media are the same. Then $n_{32} = n_{12} = 1/n_{21}$, and the above relation becomes

$$\sin \theta_3 = \sin \theta_1$$

or

$$\theta_3 = \theta_1$$

The emerging ray is in exactly the same direction as the incident ray, although it is displaced relative to the incident ray by an amount that depends on the thickness of the central plate and its refractive index, as in Fig. 17-14.

If we imagine that the central region becomes vanishingly thin, we write Snell's law in the form

$$\frac{\sin \theta_1}{\sin \theta_3} = n_{31}$$

But from the result given above we have

$$\frac{\sin \theta_1}{\sin \theta_3} = n_{32} n_{21} = \frac{n_{32}}{n_{12}}$$

Thus

$$n_{31} = \frac{n_{32}}{n_{12}}$$

FIGURE 17-14 (*From PSSC "Physics," D. C. Heath and Company, Boston, 1965.*)

If the central region were a vacuum, this relation could be written in still simpler form as

$$n_{31} = \frac{n_3}{n_1}$$ [17-4]

In words, *the index of refraction of one medium relative to another is simply the ratio of their respective indices of refraction relative to vacuum.*

When a ray is incident from a rarer medium to a denser medium, as in Fig. 17-10a, the transmitted ray is bent toward the normal and the angle of transmission can never be as large as 90°. In this case $n_{21} > 1$, and θ_1, the direction of the incident ray, can assume any value between 0 and 90°, while the transmitted ray makes an angle with the normal that ranges from 0° up to $\theta_2 = \sin^{-1}(1/n_{21})$, which is always less than 90° for $n_{21} > 1$. But when a ray is incident from

FIGURE 17-15 Six light beams enter the prism from the left. The six transmitted beams strike the glass-air interface along the hypothenuse at six different angles of incidence, the angle increasing steadily from the top to the bottom of the picture. Observe that the two lowest beams are totally reflected back into the glass. *(From PSSC "Physics," D. C. Heath and Company, Boston, 1965.)*

a denser into a rarer medium, the refracted ray is bent away from the normal and can emerge at an angle of 90°. What if the angle of incidence in the dense medium is made even greater than that for refraction at 90°? Experiment then shows that *all the light is reflected;* none is refracted. This might have been guessed from Snell's law. Because the sine cannot take on a value that is greater than 1, [17-2] has no solution for $\sin \theta_2$ if θ_1 is so large that $(\sin \theta_1)/n_{21} > 1$ (remember, in this situation $n_{21} < 1$). If there is no solution for θ_2, this *may* mean that no light is transmitted; and if no light is transmitted, the incident light must all be reflected at the surface. This

Total internal reflection

effect, known as *total internal reflection,* occurs when the angle of incidence exceeds the critical value $\theta_c = \sin^{-1} n_{21}$, where it is understood that the ray is incident in the denser medium 1 and hence n_{21} is less than 1. A light beam incident in glass can be totally reflected from the glass-air surface as shown by the two lowest beams in Fig. 17-15.

17-6 REFRACTION AND THE PARTICLE THEORY

Snell's law tells us how to trace light rays as they pass from one medium to another. It is just a matter of knowing the refractive indices for the media and applying geometry. This empirical knowledge suffices for designing lenses and devising simple optical instruments, but a deeper understanding of refraction would result if Snell's law could be deduced from the particle theory of light.

We follow the arguments of Newton, whose contributions to the study of light preceded his great accomplishments in mathematics and mechanics. So preeminent was Newton in his time that his views on the nature of light (he was an advocate of the particle theory) strongly influenced—indeed inhibited—those of other scientists.

According to Newton, when a beam of light goes through water or some other transparent homogeneous medium, the light corpuscles are symmetrically surrounded by molecules of the medium. Thus the medium exerts no *net* force on the light particles, which travel through it at a constant speed in a straight line. The only time a particle of light is subject to a resultant force is when it is close to the interface between two different media. Consider Fig. 17-16, where a ray of light goes from medium 1 into a denser medium 2. It seems reasonable to suppose that particles of light are attracted, rather than repelled, by nearby molecules of a medium and that this attractive force is the greater the denser the medium. Thus, as a particle approaches the dividing surface between the two media, it feels a net force toward medium 2 which speeds it up in the direction normal to the interface. Once inside medium 2, however, it is again in motion at a constant but higher speed. We can, in fact, relate the particle speeds v_1 and v_2 in media 1 and 2, respectively, to the angles θ_1 and θ_2 (see Fig. 17-17). It is clear that

$$\sin \theta_1 = \frac{v_{1t}}{v_1} \qquad \text{and} \qquad \sin \theta_2 = \frac{v_{2t}}{v_2}$$

where v_{1t} and v_{2t} are the components of the velocities tangent to the surface. Since the resultant force acting at the surface is at right angles to the surface, the velocity components along the surface are unchanged. Therefore,

$$v_{1t} = v_{2t}$$

Taking the ratio of the two sines and using this equality, we have

$$\frac{\sin \theta_1}{\sin \theta_2} = \frac{v_2}{v_1} \qquad\qquad [17\text{-}5]$$

The ratio of the sines *is* a constant for a given pair of materials. Thus, Snell's law can be obtained from the particle model, which furthermore shows that the relative refractive index n_{21} should be the constant ratio of the speeds v_2/v_1 in the two media. This was indeed a triumph for the Newtonian theory.

Since the index of refraction of a dense medium is greater than 1, the relation $n_{21} = v_2/v_1$ implies that the speed of light through

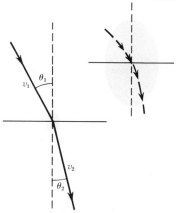

FIGURE 17-16 Newton suggested that light corpuscles near an interface between two media might experience a net force toward the medium of higher density. When moving toward the denser medium, they would be deflected toward the normal to the interface as described by the law of refraction. They also move with increasing speed as they cross the interface.

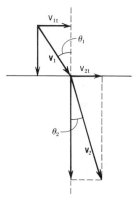

FIGURE 17-17 According to a particle theory of light, a particle in the incident beam moves with speed v_1 in a direction making an angle θ_1 with the normal to the interface, but once the particle has moved into the second medium, it has speed v_2 and moves in a direction making an angle θ_2 with the normal. As the particle crosses the interface, it experiences no net force *tangential* to the surface, and the tangential components v_{1t} and v_{2t} of the two velocities are equal. Only the normal components of velocity change.

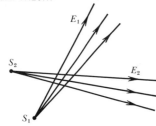

FIGURE 17-18 When two light beams from separate sources S_1 and S_2 intersect, there is no effect on the images received at the eyes of separate receivers, E_1 and E_2. This is difficult to reconcile with the particle theories of light because particles from the separate sources would be expected to collide with each other.

The particle model is inconsistent with the measured refractive indices.

a dense medium *exceeds* 3×10^8 m/s, the speed of light through air. For example, it says that the speed of light through water is *greater* than the speed through air by the factor 1.33, the refractive index for water. This prediction remained untested until 1862, when the French physicist Jean Foucault (1819–1868) measured the speed of light through water and found that it differed from the speed in air by exactly the factor 1.33. But it went the wrong way! The measured speed of light through water was that much *smaller* than 3×10^8 m/s. This experimental result leaves us no alternative but to reject Newton's particle explanation of refraction.

Actually, the very facts of vision put the naïve particle model in trouble. In the situation shown in Fig. 17-18 one light source S_1 is seen by eye E_1 while a second light source S_2 is seen by eye E_2. The two beams of lights cross; yet what each eye sees depends only on the light from the diagonal source. It is not influenced in any way by whether the other source is turned on or off. From the point of view of the particle theory this is an extraordinary state of affairs: evidently the particles in the two intersecting light beams never collide or influence each other!

17-7 NEWTON'S RINGS AND YOUNG'S EXPERIMENT

Clearly, there is something radically wrong with the particle hypothesis, at least in its classical form. Historically, it was another type of phenomenon, one characterized by patterns of alternating light and shadow, which both demanded and yielded a new theory of light.

Ironically, the first such effect to be studied has come to bear Newton's name. When a thin lens with one slightly curved spherical surface and one flat surface is placed on top of a flat plate of glass, a person looking from above sees a number of concentric bright and dark rings. The display, called *Newton's rings,* is shown in Fig. 17-19. There is a dark spot at the center when the lens and flat plate are illuminated from above and the observer views the reflected light, but when the observer views light transmitted through the lens and plate, the dark and bright regions are exactly interchanged, with a bright spot now forming the center of the pattern. Newton included an account of the phenomenon in a letter of February 6, 1672, his first scientific publication. His attempt to use the particle theory to explain the rings was rather contrived, containing many ad hoc assumptions; we shall not pursue it.

Even more baffling to the particle theory was a demonstration conceived by the English doctor Thomas Young (1773–1829). The

Side view

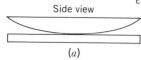

(a)

FIGURE 17-19 (a) A thin lens placed on top of a flat glass plate for Newton's rings experiment. (b) Newton's rings viewed with reflected light show darkness at the center (*Bausch & Lomb*); with transmitted light there is a bright spot at the center.

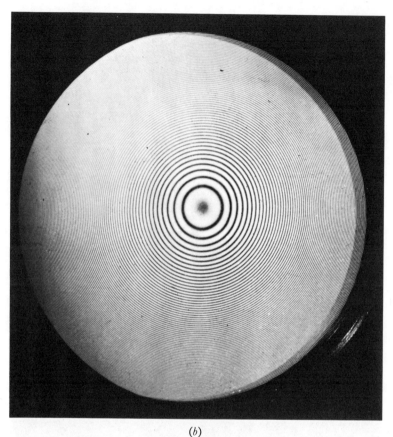

(b)

arrangement for this famous experiment is shown in Fig. 17-20. A narrow line source of light S_0 illuminates two very narrow slits S_1 and S_2 parallel to S_0. The slits are symmetrically located on either side of the perpendicular drawn from the source to the plane of the slits. The particle theory predicts two bands of light, A and B, on a screen placed beyond the slits S_1 and S_2 and total darkness at the midpoint C between A and B. What Young actually found was alternate regions of brightness and darkness, with a *bright* band exactly at the position C. In terms of a particle theory, this is absurd! It is as though a gunner were firing bullets in all directions from position S_0 and yet somehow it was not safe to stand right in the middle of the shadow cast by the opaque wall between two doors S_1 and S_2!

As a student at Cambridge Young had been known as Phenomenon Young because he was reported to know everything. Among his

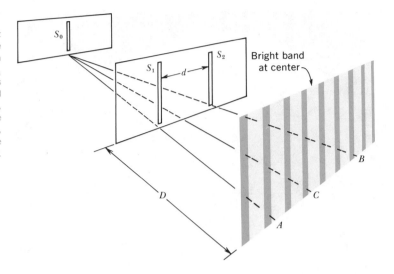

FIGURE 17-20 Young's double-slit experiment: light from a distant line source S_0 is incident nearly normally upon a screen having two closely spaced parallel slits S_1 and S_2; the light from the two parallel slits is observed to form alternating bright and dark bands (fringes) on a distant screen, with a bright fringe at the center C of the screen. According to the particle theory, there should be bright light only at the positions A and B but not at C.

interests was the study of the human ear and sound waves. Therefore it is not surprising that when he learned about the puzzling rings and bands produced by light, he sensed a similarity between that effect and the interference of sound waves or water waves. Such waves, under certain conditions such as those shown in Fig. 17-21, exhibit alternating regions of maximum and minimum disturbance.

FIGURE 17-21 Circular waves on the surface of water. (*Ken Kay, "Light and Vision,"* © *1966, Time, Inc.*)

Indeed, Young devised the double-slit experiment precisely to demonstrate similar wave properties for light. We shall study these properties in the next chapter.

SUMMARY

When a ray of light is incident upon the interface between two transparent materials, (1) the reflected ray lies in the plane of incidence (the plane containing the incident ray and the normal to the surface) and the angle of incidence equals the angle of reflection

$$\theta = \theta'$$ [17-1]

and (2) the refracted ray lies in the plane of incidence, and the ratio of the sine of the angle of incidence and the sine of the angle of refraction is a constant called the index of refraction of the second material relative to the first

$$\frac{\sin \theta_1}{\sin \theta_2} = n_{21}$$ [17-2]

This index of refraction can also be expressed in terms of the individual indices of refraction of each material relative to vacuum

$$n_{21} = \frac{n_2}{n_1}$$ [17-4]

The speed of a light wave in vacuum is a constant independent of the light intensity

$$c = 3.0 \times 10^8 \text{ m/s}$$

PROBLEMS

17-1

A monkey stands in front of a mirror twirling a banana on the end of a rope in a vertical plane that is parallel to the plane of the mirror. He looks directly at the banana and sees it rotating clockwise. In what sense does the monkey's mirror image see the image of the banana rotating?

17-2

The figure shows a luminous object O observed by an eye near two plane mirrors at right angles to each other. The observer's eye E can see the object directly, or he can see its image by reflection from one, or the other, or both mirrors. (a) Apply the laws of reflection to locate the *three* images. Note that the image arising from reflection in both mirrors is not inverted left-right; a person viewing his image in such a pair of mirror sees himself as others see him, without the left-right reversal that occurs in single plane

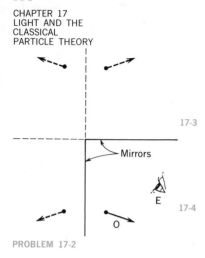

Mirrors

E

O

PROBLEM 17-2

[1] A corner reflector consists of *three* mutually perpendicular plane mirrors, like one corner of a mirrored room. It can be shown that a ray of light incident upon a corner reflector is always reflected back in a direction opposite to the incident ray, whatever the particular origin of the incident ray. A collection of 100 such corner reflectors (prisms of fired silica) was placed on the moon by the astronauts in the Apollo 11 mission. A beam of light from Earth incident upon the corner reflectors is reflected back so that it can be seen directed at the point of origin. Measuring the roundtrip flight time permits the Earth-moon distance to be determined with great precision.

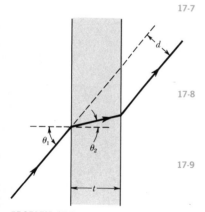

PROBLEM 17-7

mirrors. (*b*) Suppose that parallel rays of light are directed at or near the intersection of two perpendicular plane mirrors. Show that the rays reflected from both mirrors emerge in the direction opposite to that of the incident rays.[1]

17-3   It takes 20 s for the temperature of a small black object to rise by 2°C when the object is 8.0 m from a bright point source of light. How long does it take for the temperature to rise by the same amount when the small black object is 2.0 m from the same point source (assuming that the object has the same orientation relative to the incident rays of light)?

17-4   A camera is properly set for taking a picture; the exposure time is $\frac{1}{50}$ s, and the lens opening is $f/8$ (the setting $f/8$ means that the diameter of the lens opening is one-eighth of the distance between the lens and film, or the focal length f). If the exposure time is reduced to $\frac{1}{200}$ s, what setting for the lens aperture will allow the same amount of light to reach the film?

17-5   The horizontal bottom of a water tank is made of glass. A light beam incident from the air above the water strikes the water making a 55° angle with the outward normal to the water surface. (*a*) What angle does the transmitted beam make with the inward normal? (*b*) What angle does the light beam make with this same normal direction after it enters the glass (refractive index, 1.50)?

17-6   A person standing close to a reflecting convex surface sees his image reduced, while a person standing close to a reflecting concave surface sees his image enlarged. Explain this behavior qualitatively.

17-7   A ray of light is incident upon a glass plate of thickness t. The transmitted ray on the opposite side of the plate is displaced from the projected path of the incident ray by an amount d, as shown in the figure. (Assume vacuum on both sides of the plate.) Derive an expression for d in terms of t and the angles of incidence and refraction, θ_1 and θ_2, at the first surface.

17-8   (*a*) The index of refraction of water relative to air is 1.33. A light ray *in water* is incident upon a water-air surface. If the angle of incidence is 30°, what is the *sine* of the angle of transmission into the air? (*b*) What happens for an angle of incidence of 60°?

17-9   Suppose that an observer is situated at the surface of a liquid having an index of refraction of $\frac{5}{3}$. He is directly above a small light source at the bottom of the tank, 4 ft from the surface (see the figure). What is the apparent depth of the light source for this observer. *Hint:* Consider at least two rays from the light source, apply Snell's law, and assume that the angles of the light rays with the normal to the liquid-air interface are small.

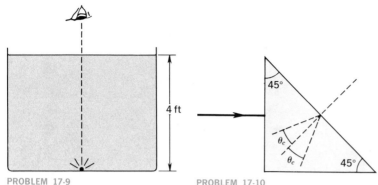

PROBLEM 17-9   PROBLEM 17-10

17-10 Explain how it is possible to reflect a beam of light totally by using a prism (see the figure). (Totally reflecting prisms are found to absorb less light energy from an incident beam than a mirror does.)

17-11 (*a*) What is the critical angle for total internal reflection at a glass-water interface when the index of refraction of the glass is 1.6? (*b*) In which material, water or glass, does total internal reflection take place?

17-12 (*a*) What is the critical angle for the total internal reflection of light within diamond (index of refraction = 2.42)? (*b*) How is the high refractive index for diamond related to the fact that gems of diamond, with a number of facets at relatively small angles between their surfaces, are particularly brilliant when exposed to light?

17-13 A large tank filled with carbon disulfide to a depth of 20.0 cm has a point light source at its bottom center. Through how large an area of the carbon disulfide–air interface will light from this source pass? (The index of refraction of carbon disulfide is 1.63.)

17-14 The figure shows a bent cylinder of plastic having a high index of refraction. A ray of light is incident upon the cylinder in a direction slightly off the axis of the cylinder. Show that if the bends in the plastic tube are fairly gentle, the incident ray of light is repeatedly reflected and remains trapped in the plastic because of total internal reflection. A device of this sort is commonly known as a *light pipe*.

PROBLEM 17-14

THE WAVE THEORY AND INTERFERENCE OF LIGHT

The alternating regions of brightness and darkness in Young's experiment are readily explained in terms of wave behavior. To familiarize ourselves with the properties of waves, we shall first examine the motion of several simple mechanical systems. Application of nothing more than momentum and energy conservation will reveal the basic features of wave disturbances.

18-1 WAVE PROPAGATION ALONG A LINE OF COUPLED MASSES

Something of what a wave is (and what it isn't) can be gathered from Fig. 18-1. An uncoupled series of identical elastic masses is equally spaced along a line. The objects might be gliders on a frictionless air track or steel balls on a smooth lane. If the leftmost mass is hit sharply toward the right by a blow, it will acquire a constant speed v_0 to the right before it has been displaced appreciably from its initial position. We know what happens next from our earlier discussions of collisions: the mass m_1 collides elastically with m_2, to which it transfers all its momentum and kinetic energy, itself coming to rest. A whole series of identical collisions follows, until at last the rightmost mass is found to be traveling with the same momentum and kinetic energy that the mass at the extreme left started out with.

This sequence of elastic collisions is thus characterized by the long-distance transfer of momentum and kinetic energy. The speed at which this transfer is accomplished is v_0, the speed of each moving particle. Therefore for this special process, unlike genuine wave motion, the greater the amounts of momentum and kinetic energy transferred (mv_0 and $\frac{1}{2}mv_0{}^2$, respectively), the greater the speed at which they move. Imagine the problems of an orchestra conductor if the louder an instrument was played the sooner it was heard!

If the lined-up masses are interconnected by identical springs, as in Fig. 18-2, genuine wave motion becomes possible. Let us assume that

FIGURE 18-1 A disturbance transmitted along a line of identical elastic objects. One object initially at rest is hit by its left neighbor. The left neighbor comes to rest while the object that has been hit moves on to hit its neighbor, and so on. Finally the disturbance has moved a long distance but the individual objects have moved no farther than the distance to a nearest neighbor.

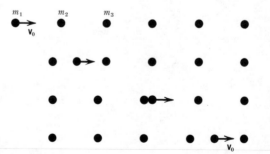

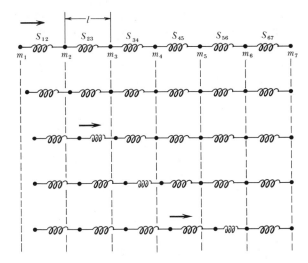

FIGURE 18-2 A disturbance transmitted along a line of identical objects coupled by elastic springs. Again the distubance moves over long distances, but now the individual objects barely move from their initial positions.

the springs are of negligible mass and that at first each spring is neither stretched nor compressed. Again we imagine the mass on the left end to be struck a sharp blow and to be set quickly in motion to the right with the initial speed v_0. But as soon as m_1 moves to the right, it compresses spring S_{12}, which exerts a force to the left on m_1, thereby slowing it. At the same time the spring produces a force of equal magnitude to the right on mass m_2, setting it in motion. Thus, there is a soft collision between the coupled masses m_1 and m_2. As m_1 is brought to rest by the spring S_{12}, m_2 is set in motion to the right; and again the momentum and kinetic energy originally carried by m_1 are transferred to m_2. The process continues as each particle in turn is set in motion and then brought to rest. What distinguishes the process from that of Fig. 18-1 is that the energy and momentum travel along the chain of coupled masses *not* at speed v_0 but at a different (higher) speed *which depends only on the mass of each particle and the stiffness of the springs.*

The speed of momentum and energy transfer along the chain depends only on the mass of each object and the stiffness of the springs.

Accepting, for the moment, this remarkable result as experimentally determined, we have encountered the two fundamental characteristics of waves in general: (1) in a wave, energy and momentum are transported over large distances, yet individual elements of the medium undergo only small displacements, and (2) the speed of a wave (the speed at which energy and momentum are propagated) depends only on the properties of the medium and is independent of the source and of the amounts of energy and momentum transmitted.

Let us emphasize one difference between the transport of energy and of momentum. Any wave carries a positive amount of energy along with it, but the amount of momentum carried in the direction of wave propagation may be negative or even zero. That is, the traveling momentum *vector* may even be pointing in the direction opposite to its own motion. For example, if right is chosen as the positive direction in Fig. 18-2, we can say that positive momentum has been transferred in the positive direction. But if mass m_1 is initially displaced to the left, rather than to the right, then spring S_{12} is stretched, rather than compressed, and m_2 is pulled to the left. The momentum that is transferred from particle to particle is momentum to the left, negative momentum, and it is always transferred to the next particle on the right. Therefore in this wave negative momentum is being transferred in the positive direction. Figure 18-3 shows graphs of these two waves. The displacement x of each mass from its initial position is plotted vertically, with the positive direction upward. The horizontal coordinate X gives the location of each particle along the chain.

Both positive and negative momentum can be transferred along the chain.

Finally, suppose that m_1 is first struck toward the right and then a little later struck toward the left, bringing it back to its initial position. A wave pulse is generated which travels to the right as shown in Fig. 18-4. The leading edge represents momentum to the right (positive momentum) and the trailing edge momentum to the left (negative momentum) in the same magnitude. Here the net momentum transfer is zero. After the pulse has traveled through the chain, each mass is back at its equilibrium position.

EXAMPLE 18-1   Suppose that a mass near the center of the chain is suddenly set in motion toward the right. What sort of wave disturbance is generated?

SOLUTION   For definiteness we imagine the mass initially set in motion to the right to be m_6. Then spring S_{67} is compressed while spring S_{56} is stretched (see Fig. 18-5). Both the adjoining masses, m_5 and m_7, are set in motion to the right, but the mass m_7 generates a wave to the right, while mass m_5 generates a wave to the left. Although the two waves travel in opposite directions, in both waves the momentum transferred from mass to mass is positive (to the right).

+   18-2   MATHEMATICAL
DERIVATION OF THE
WAVE SPEED

With the help of some simplifying assumptions we can substantiate our assertion that the wave speed is independent of the momentum and energy being transferred. It was said earlier that mass m_2 is set

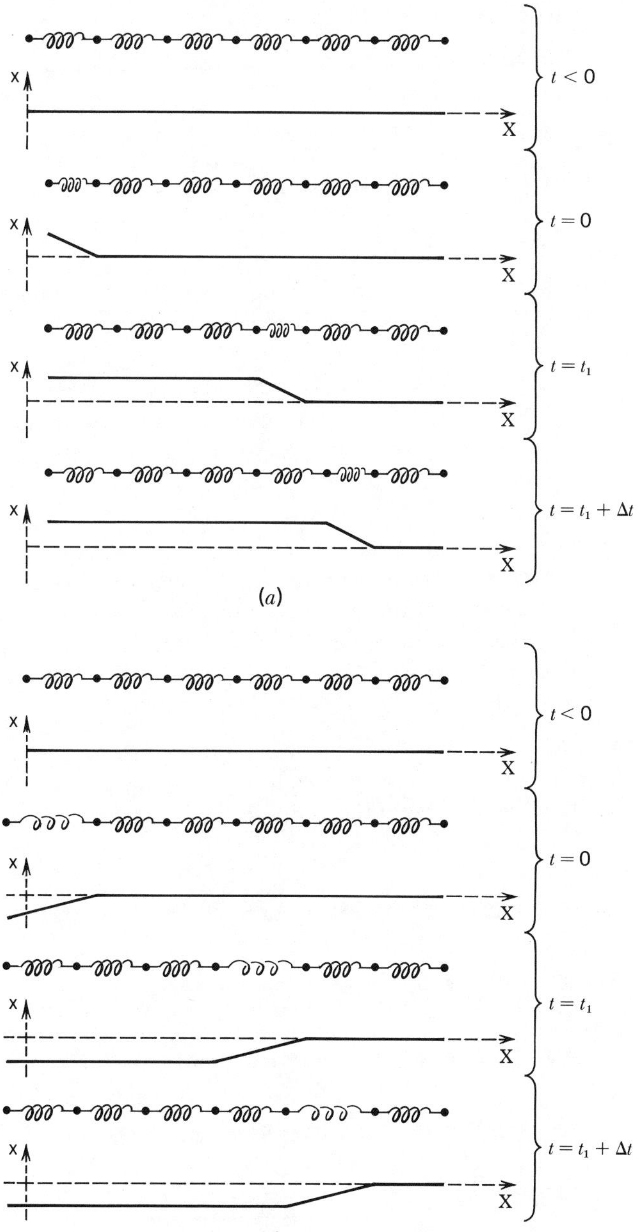

FIGURE 18-3 The displacement x of the objects in Fig. 18-2 from their initial positions as a function of object position X along the line. (a) The object at the left end is displaced to the right at $t = 0$. (b) The object at the left end is displaced to the left at $t = 0$.

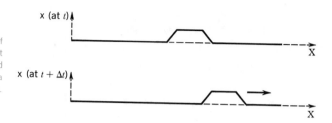

FIGURE 18-4 Similar plots to those of Fig. 18-3 except that the object at the left end was displaced to the right at $t = 0$ and then displaced back to its initial position a short time later.

into motion when spring S_{12} is compressed. Strictly speaking, as soon as spring S_{12} is deformed, m_2 begins to move and thereby deforms spring S_{23}. This implies that mass m_3, and still other masses to a much lesser degree, are in fact immediately affected by the motion of m_1. However, we suppose that when masses m_1 and m_2 interact through spring S_{12}, no other masses and springs are noticeably affected, so that these two masses and their connecting spring can be treated as an isolated system. Moreover, we shall take the maximum displacement of m_1 (and later of m_2) to be small compared to the length L_0 of the relaxed spring. Thus the wave speed, the speed with which energy and momentum are transferred, is simply the distance L_0 through which the disturbance advances in going from m_1 to m_2 divided by the time Δt_L required for this advance. Everything now depends on determining Δt_L.

It will be convenient to discuss the events shown in Fig. 18-6 from a reference frame at rest relative to the system's center of mass. We know that in this frame the collision between the equal masses m_1

FIGURE 18-5

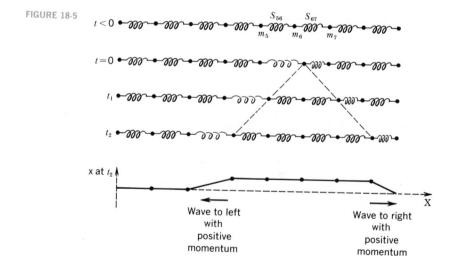

and m_2 becomes symmetrical as shown in Fig. 18-7. Initially each mass approaches the center of mass at speed $\frac{1}{2}v_0$ (v_0 being the initial speed of m_1 in the laboratory system); they come to rest; then they reverse their previous motion. The time elapsing between the instant when m_1 moves with the speed $\frac{1}{2}v_0$ to the right and the later instant when m_1 moves left with the same speed is just the interval Δt_L.

The motion of each mass is controlled by the force of the spring, which is given for small changes in length ΔL by $F = k\,\Delta L$, where k is the stiffness constant of the spring. Imagine the x axis along the line of the motion with its origin at the *initial* position of the left particle in Fig. 18-7. From our definition of force we have

$$a = \frac{F}{m} = \frac{-k\,\Delta L}{m} = -\frac{2k}{m}x \qquad [18\text{-}1]$$

where x and a are the position and acceleration of the left particle. (Note that in this reference frame both ends of the spring move equal amounts but in opposite direction; therefore the displacement Δx of the left end of the spring, and hence the position x of the left particle, is related to the total change in length of the spring ΔL by $\Delta L = 2\Delta x = 2x$.)

Equation [18-1] can be solved for the displacement (see Problems 8-19 and 8-20) to give

$$x = \frac{1}{2}v_0\sqrt{\frac{m}{2k}}\sin\left(\sqrt{\frac{2k}{m}}\,t\right) \qquad [18\text{-}2]$$

We can quickly check that [18-2] gives the correct initial position and velocity. For $t = 0$, we have x $= 0$, as required. And the initial velocity should be equal to the average velocity over a very short time interval Δt at the beginning. But for very small values of t, the sine function can be replaced by its angle. Thus this initial average velocity is

$$\langle v \rangle = \frac{\Delta x}{\Delta t} = \frac{1}{2}v_0\sqrt{\frac{m}{2k}}\sqrt{\frac{2k}{m}}\frac{\Delta t}{\Delta t} = \frac{1}{2}v_0$$

just as required.

The speed is $\frac{1}{2}v_0$ when $x = 0$ at $t = 0$; it is again $\frac{1}{2}v_0$ when x is again zero at $t = \Delta t_L$. Thus

$$\sin\left(\sqrt{\frac{2k}{m}}\Delta t_L\right) = \sin 180° = \sin \pi$$

$$\Delta t_L = \pi\sqrt{\frac{m}{2k}}$$

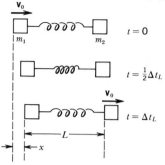

FIGURE 18-6 Two objects of identical mass are connected by an elastic spring of length L. Object 1 is projected to the right with initial speed v_0; the interaction force transmitted by the spring sets object 2 in motion, and it acquires the speed v_0 just as the first object comes to rest.

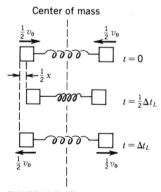

Center of mass

FIGURE 18-7 The motion of the two objects of Fig. 18-6 as viewed from a center-of-mass reference frame, a frame moving to the right with speed $\frac{1}{2}v_0$ with respect to the reference frame of Fig. 18-6.

Then returning to the original reference frame of Fig. 18-6, we find that the wave speed v is given by

$$v = \frac{\Delta X}{\Delta t} = L_0 \frac{\sqrt{2}}{\pi} \sqrt{\frac{k}{m}}$$

As v_0 does not enter the expression for v, the wave speed does not depend on how fast the individual elements of the chain move. The wave speed is therefore independent of the momentum and energy transported but dependent on the properties L_0, m, and k of the medium. This is what we sought to prove.

Actually, our formula for the wave speed is not quite right. Since the chain beyond m_2 does in fact receive some momentum before m_1 returns to equilibrium, the wave speed is greater than we computed above. A more detailed analysis shows the correct value to be

[18-3] $$v = L_0 \sqrt{\frac{k}{m}}$$

or about twice that given by our simplified calculation. In any event, L_0, m, and k enter both relations in the correct fashion. As our intuition would tell us, when the springs are made stiffer, i.e., when k is increased, the wave speed increases. Moreover, when m, the mass of the individual elements, is increased, their motions become more sluggish and so the wave speed decreases.

18-3 OTHER TYPES OF WAVES

Longitudinal wave: displacement along the chain; transverse wave: displacement at right angles to the direction of wave propagation

So far we have considered only waves in which the mass elements move along the line of energy flow. These waves are called *longitudinal*. However, if the first mass of a horizontal chain is suddenly displaced vertically, another type of wave is generated. The remaining masses consecutively undergo the same vertical motion as the first, because of the forces in the stretched springs (see Fig. 18-8). Again energy and momentum are transferred down the chain, but now the momentum of each mass lies athwart the direction of the wave. Such a wave is called *transverse*, a familiar example being the wave produced when a taut string is pulled aside at one of its points and then released. The string may be regarded as an infinite collection of coupled masses, with the tension playing the role of springs. As before, the speed with which the waveshape moves depends only on the properties of the string and not on the crosswise velocities of the particles.

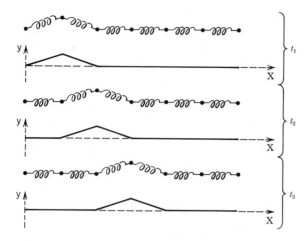

FIGURE 18-8 A transverse-wave disturbance transmitted along a line of objects coupled by elastic springs; the direction of the momentum transmitted is now perpendicular to the direction of energy transmission along the line.

In solid and liquid media more complex waves are possible, in which the disturbance may propagate in two or three dimensions and be neither wholly longitudinal nor wholly transverse. (In ordinary ocean waves the water particles execute ellipses around their equilibrium positions.) But all these mechanical waves are essentially alike: the motion of masses at one point in a medium affects the motion of adjoining masses, so that a disturbance progressively advances.

What happens when a wave traveling in one direction meets a wave traveling in the opposite direction, say two wave pulses along a string? When the two disturbances traveling as shown in Fig. 18-9 reach the same point simultaneously, the net disturbance there is merely the algebraic sum of the separate pulses; at later times the two pulses are found continuing in motion to the left and right with no change in shape or speed whatsoever. We can summarize by

18-4 SUPERPOSITION OF WAVES

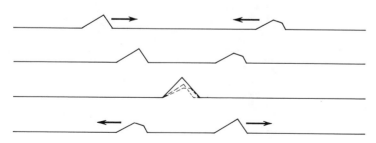

FIGURE 18-9 Two disturbances of different shapes traveling in opposite directions along a stretched string. When the two disturbances meet, the net disturbance is simply the sum, or superposition, of the two individual disturbances.

Superpose individual displacements
algebraically to find the net displacement.

Interference: the superposition of individual
waves

Constructive and destructive interference

saying that at every location and at every instant the net displacement of the string is the sum, or *superposition*, of the displacements separately produced by the individual waves. This principle of superposition holds for the displacement or momentum in any compound wave, provided the component disturbances are not too large.

The combination of waves by superposition is called *interference* (not a very apt name for a process whereby two waves may pass through each other undeformed, but we are stuck with it). If, as in Fig. 18-9, the net disturbance is *larger* than either component disturbance, the interference is said to be *constructive*. Figure 18-10 shows a case of *destructive* interference. There the one wave is *up* while the other *wave* is *down* (we say the two waves differ in their *polarity*), so that superposition gives a net disturbance *smaller* than either wave separately.

A remarkable kind of interference is seen when two waves in opposite directions have the same shape, apart from a left-right reversal and an up-down reversal (see Fig. 18-11). There is one instant when the string shows no displacement at all, but after this moment of complete destructive interference, each wave emerges, unchanged in shape, to travel as if it had not encountered the other.

This consequence of superposition can be made less surprising by examining a similar situation involving two longitudinal disturbances. Imagine, in the line of masses shown in Fig. 18-1, a second disturbance of negative momentum traveling to the left. It was started at the right end of the chain at the same time the disturbance of positive momentum traveling to the right was started at the left. When the two waves meet at the center, two particles of equal mass moving with equal speeds in opposite directions collide. The collision being perfectly elastic, the conservation laws demand the reversal of the velocities of the two particles (see Example 8-2). This is precisely as though the negative momentum originally traveling to the left kept traveling to the left while the positive momentum

FIGURE 18-10 Two disturbances as in Fig. 18-9, only now the two have opposite signs. The observed result when the two meet shows that the superposition involves an *algebraic* sum.

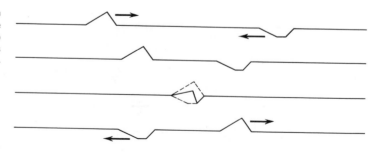

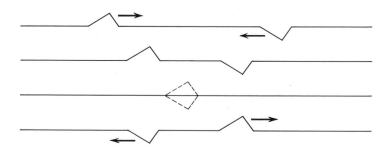

FIGURE 18-11 Two disturbances as in Fig. 18-10, only now the shapes of the disturbances of opposite sign are such that there is complete cancellation when they meet, i.e., completely destructive interference.

originally traveling to right kept traveling to the right. The disturbances "go right through" each other. Only at the instant of the central collision are all the particles at rest. Then all the energy of the system is stored as potential energy of deformation.

If one wave is oblivious of the presence of another, how does it react to the presence of a boundary? Suppose, for instance, that the last spring in a chain of coupled masses is attached to a solid wall, i.e., an effectively infinite mass, as shown in Fig. 18-12. A wave pulse is generated at the left end of chain, each mass moving in turn to

18-5 REFLECTION OF WAVES

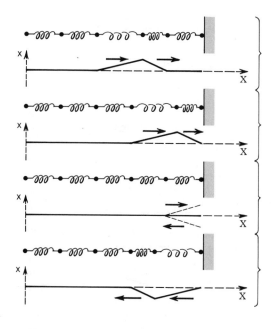

FIGURE 18-12 A positive longitudinal disturbance moves to the right along a line of objects coupled by elastic springs. The last object at the right is fixed to a massive wall, and the positive disturbance is reflected to the left as a negative disturbance. At one instant at the middle of the collision with the massive wall there is complete destructive interference between the back half of the positive disturbance moving to the right and the front half of the negative disturbance reflected to the left.

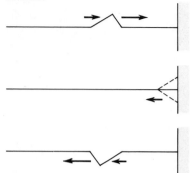

FIGURE 18-13 A positive transverse disturbance on a stretched string when incident upon an end of the string tied securely to a massive wall is reflected as a negative disturbance. As in Fig. 18-12, at one instant at the middle of the collision with the wall the back half of the positive incident disturbance interferes completely destructively with the front half of the reflected disturbance, and at this instant the string is undisplaced.

the right and each spring undergoing in turn the same compression followed by a restoration to its natural length. But the last spring, when compressed by the mass on its left, cannot move still another mass to the right, as the other springs do. Consequently, it is *compressed more* than the others, and it exerts greater force back on the mass to its left. That mass, then, is not merely brought to rest but is set in motion back to the left, as are all the other masses in turn. In short, the wave's momentum is first to the right, but after the wave encounters the solid wall, the wave is reflected to travel to the left with momentum that is also to the left.

Exactly the same sort of behavior is shown by a wave along a taut string attached rigidly to a wall (see Fig. 18-13). Here it is the inability of the string to pull the wall in the transverse direction that accounts for the reflected momentum of opposite sign. Since momentum *up* becomes momentum *down*, the wave returns with reversed polarity.

A very different sort of boundary is exhibited by a chain of coupled masses of which the last mass is free to its right (see Fig. 18-14). When the wave reaches it, this last mass overshoots, thereby *stretching* the last spring *more* than the other springs. Consequently, a reflected wave is propagated to the left, but now the reflected wave has the *same*

FIGURE 18-14 A positive disturbance moving to the right along a line of objects coupled by elastic springs is reflected from a free end at the right. Since the last object has no spring or object on its right to slow it down, it moves twice as far as the other objects until it is finally stopped by the restraining force from the spring to its left. But this extra displacement of the last object to the right initiates a reflected positive wave to the left. While the disturbance is being reflected from the loose end, the back part of the incident disturbance moving to the right interferes constructively with the front part of the reflected disturbance moving to the left.

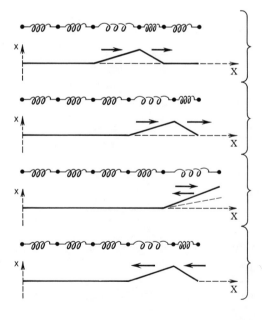

polarity as the incident wave. Similarly for the reflection from the free end of a string (Fig. 18-15).

Although these two cases, the perfectly hard boundary and the perfectly soft boundary, are the extremes, from them we can infer what happens when a wave traveling in one medium encounters a second medium having different properties and therefore a different wave speed. For example, one might have one chain of coupled masses attached to a second chain in which the individual mass elements are heavier. We know then that the wave will travel slower in the denser medium, following [18-3]. When a wave approaches the boundary from the low-density side (at the left in Fig. 18-16), it is partly transmitted into the denser medium and partly reflected at the boundary to travel to the left. The transmitted wave travels at a lower speed to the right without a sign or polarity reversal, while the reflected wave travels back at the same speed but with a reversal. This reversal is to be expected, since the incident wave meets a more massive, or harder, medium. The reverse situation is shown in Fig. 18-17, where a wave encounters the boundary, approaching it from the denser medium. Here we expect, and indeed find, that neither the reflected nor the transmitted wave undergoes a change in polarity.

Remember that when a beam of light strikes a transparent medium, some of the energy is transmitted and some reflected. Thus, the results of this section square with a wave theory of light. Later we shall see that light does indeed change its polarity on reflection from a denser medium.

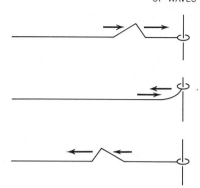

FIGURE 18-15 A positive transverse disturbance on a stretched string with one end free to slide up and down as shown is reflected from the free end as a positive disturbance.

At a hard boundary, up momentum is reflected as down momentum; at a soft boundary, up momentum is reflected as up momentum.

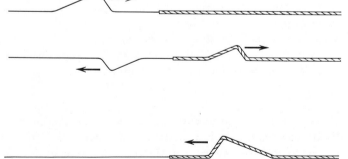

FIGURE 18-16 When a disturbance traveling along a string comes to a section having a large mass per unit length, some of the disturbance is transmitted into the new section and some is reflected back. The polarity (sign) of the transmitted disturbance is the same as that for the incident disturbance, but the polarity of the reflected part is reversed.

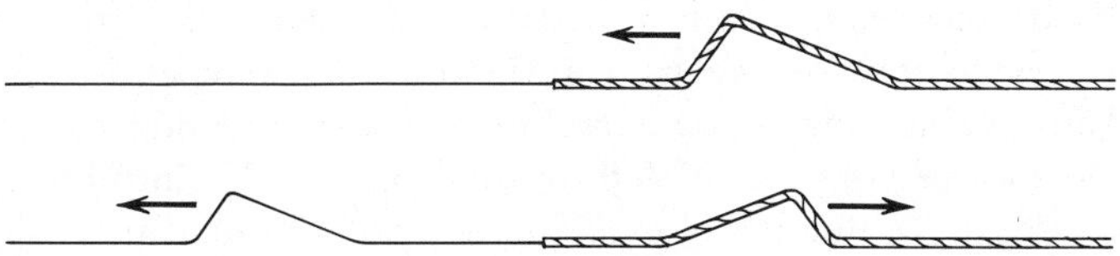

FIGURE 18-17 When a disturbance traveling along a string comes to a section having less mass per unit length, both the transmitted and reflected parts have the same polarity as the incident disturbance.

18-6   PERIODIC WAVES

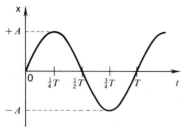

FIGURE 18-18   The time dependence of the displacement of a particle experiencing a periodic sinusoidal disturbance. The period is T and the amplitude A.

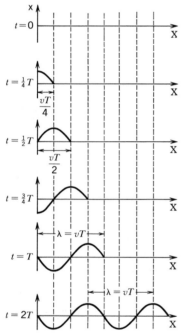

FIGURE 18-19   The progress of a sinusoidal disturbance along a line of objects coupled by springs as the object at the left end experiences the disturbance of Fig. 18-18. After one full period T, the front of the disturbance moving with speed v has progressed one full wavelength $\lambda = vT$ along the line.

The period: the time for a particle to execute one complete cycle

Wavelength: distance wave advances in one period

Till now, only special waves have been considered along a chain of coupled masses. These were sharply localized disturbances corresponding to the first mass m_1 being abruptly set into motion with speed v_0. But we haven't really restricted ourselves. If, instead, the speed of m_1 changes continuously, we can always suppose this change to have been caused by a succession of small sharp blows. Each blow sends a wave pulse down the chain, and the superposition of these pulses is the wave set up by the continuous motion of m_1.

From this way of looking at an arbitrary wave, a basic result emerges at once: *the remaining masses undergo exactly the same motion as the first, only at later times.* For, since this is true for each component pulse, it must be true for the sum of these pulses. Thus, the motion of the first mass and the wave speed completely characterize the disturbance.

A kind of wave encountered again and again in physics is generated when the first mass in the medium is made to oscillate like a sine function about its equilibrium position, as shown in Fig. 18-18. The time required to take the mass through one whole cycle, i.e., a positive excursion followed by a similar negative excursion, is designated T, the *period* of oscillation. In a time $t = T/4$, m_1 is moved a distance A from its equilibrium position. Between $t = T/4$ and $t = T/2$, m_1 is moved back to its starting point at $x = 0$. Between $t = T/2$ and $3T/4$, the mass moves a distance A on the other side of its equilibrium position, returning to $x = 0$ finally at $t = T$. The maximum magnitude of the displacement is called the *amplitude A.* As we have just seen, the other masses will successively take up this same motion. The early stages of the wave are shown in Fig. 18-19. Each part of the figure may be thought of as a snapshot, giving an instantaneous graph of the displacement x as a function of the location X. It is seen that in any length of time t, the wave's leading edge advances a distance vt, where v is the wave speed. In particular, in the time interval T, during which m_1 has executed one complete oscillation, the wavefront has advanced a distance which we call the *wavelength λ.* By this definition,

$$\lambda = vT \qquad [18\text{-}4]$$

The wavelength λ represents the *distance in space* over which the disturbance repeats itself, whereas the period T is the *interval in time* over which any one element in the disturbance repeats its motion. Thus, a sinusoidal wave is *periodic* both in time and in space. We can express [18-4] in a slightly different way by introducing the *frequency f,* the number of oscillations of any one mass per unit time. Since T is the time for one oscillation, $f = 1/T$, so that

$$v = \lambda f \qquad\qquad [18\text{-}5]$$

Sinusoidal waves are important because many common wave generators oscillate periodically in the fashion of Fig. 18-18; i.e., the displacement is a sinusoidal function of the time. But much more fundamental is the fact that *any wave disturbance, whatever its shape, can be regarded as the superposition of sinusoidal waves of various frequencies.* Although a mathematical proof of this statement is beyond the scope of this book, we shall not hesitate to take advantage of the result. There is no need to deal with arbitrary wave disturbances: if we can solve a problem for a simple sine wave, we can solve it by superposition for any shape.

Having developed the principal features of mechanical waves, we can now explain Young's results (Section 17-7), namely, the presence of a bright band in the heart of what ought (by the particle theory) to be total shadow. In what follows we shall assume that light indeed consists of waves without committing ourselves to what is "waving."

The essential parts of Young's double-slit experiment are shown again in Fig. 18-20. On the left are the two slits S_1 and S_2 (in top view), assumed to act as identical periodic sources emitting waves in all directions to the right of the plane of the slits. The slits are separated by the distance d. The screen is parallel to the plane of the slits at a distance D from it. Tranverse positions on the screen are labeled y, chosen to be zero at the central point. That point, being equidistant from the two identical sources, will receive identical

18-7 THE DOUBLE-SLIT EXPERIMENT

Young's interference experiment

FIGURE 18-20   The essential quantities used in the analysis of Young's double-slit experiment (Fig. 17-20). The separation of the parallel slits S_1 and S_2 is d, and D is the distance from the plane of the slits to the plane of the screen used to observe the fringes. The distances from a given point on the screen to the two slits are D_1 and D_2. The difference Δ between these two distances determines the degree of constructive or destructive interference at the screen. Values of the coordinate y along the screen where bright fringes exist are identified by subscripts giving the number of the fringe away from the center.

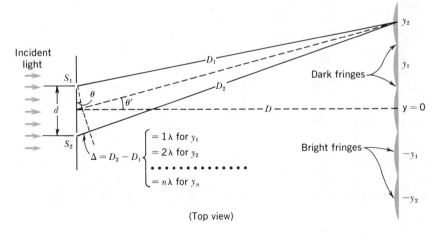

(Top view)

signals from them. With wave crests (and troughs) arriving simultaneously, the total disturbance is enhanced over that from either slit separately; therefore, the point $y = 0$ is a location of maximum disturbance or *brightness*. We have at once the reason for the central bright fringe. In addition there are other locations of maximum brightness along the screen. Let us denote the respective distances from slits S_1 and S_2 to a point on the screen by D_1 and D_2. Now if D_1 and D_2 differ by exactly 1, 2, 3, . . . wavelengths, the separate wave disturbances will again interfere constructively at the point, giving maximum brightness. Alternating with the bright spots (or parallel fringes, in two dimensions) are points for which the distances D_1 and D_2 differ by an odd number of half wavelengths, so that a crest from one source arrives simultaneously with a trough from the other source. Here the two disturbances interfere destructively,[1] and at these places the fringes are therefore dark. A photograph of a set of fringes is shown in Fig. 18-21; also see Fig. 18-22.

It is an easy matter in certain cases to write an expression giving the distance y_n of the nth bright fringe from the center of the screen. Referring to Fig. 18-20, we see that how the two waves combine is controlled by Δ, the difference between the lengths D_1 and D_2. When Δ is an integral multiple of λ, we have maximum brightness; when Δ

[1] For perfectly destructive interference it would be necessary for the two disturbances to have equal magnitudes at the screen. Although this is very nearly the case, it is not exactly so because, except for the central fringe, one disturbance must travel a little farther than the other and hence is a little weaker.

Interference is controlled by Δ, the difference in the two path lengths.

FIGURE 18-21 The fringes near the center of the screen in a Young's double-slit experiment. (*Dr. Brian J. Thompson, The Institute of Optics, University of Rochester.*)

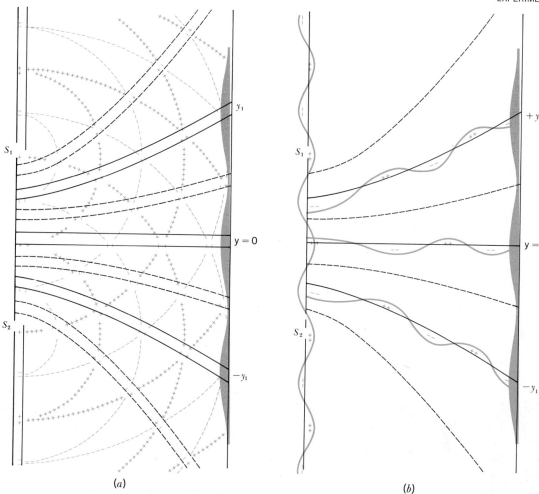

(a)   (b)

is an odd multiple of $\frac{1}{2}\lambda$, we have darkness. Thus, for the center of a bright fringe,

$$\Delta_n = n\lambda \qquad \text{where } n = 0, 1, 2, \cdots \qquad [18\text{-}6]$$

If D_1 and D_2 are much larger than d, the small triangle at the slits containing the angle θ is nearly a right triangle and we can write

$$\sin \theta \approx \frac{\Delta}{d} \qquad [18\text{-}7]$$

where θ also nearly equals θ', the angle between the line drawn from a point midway between the slits perpendicular to the screen and

FIGURE 18-22 The interference pattern in Young's double slit experiment for the region between the plane of the slits and the plane of the screen at an instant of time when the disturbance at each slit has its maximum positive value. Curves along which the disturbance is always identically zero (nodes) lie between curves along which the disturbance varies periodically with maximum amplitude.

a line drawn from this same midpoint to the point on the screen where the light intensity is being observed.

Moreover, *if the ratio of the distance y* (between the center of screen and the observation point) *to the distance D* (between the plane of the slits and the plane of the screen) *is small,* then the angle $\theta' \approx \theta$ is small and we write

[18-8] $$\sin \theta \approx \theta \approx \theta' \approx \frac{y}{D}$$

Here, of course, we have expressed θ' in radians.

For these special conditions, the position of the *n*th bright fringe is given by

$$\frac{y_n}{D} \approx \frac{\Delta_n}{d}$$

or, using [18-6]

[18-9] $$y_n \approx n\lambda \frac{D}{d}$$

Equation [18-9] shows that for distant observation points and small angles the bright fringes are equally spaced $\lambda D/d$ apart. Moreover, if D, d, and y_n are measured, λ can be computed. Young found λ to be roughly 5×10^{-7} m. Compared to the dimensions of ordinary objects, the wavelength of light is exceedingly small. It is often convenient to use the angstrom (Å), equal to 10^{-10} m; thus Young found that the wavelength of light is roughly 5000 Å.

The angstrom (Å) unit, 10^{-10} m

Using Young's result for the approximate wavelength of light, it is possible to compute the approximate frequency of light

$$f = \frac{1}{T} = \frac{v}{\lambda} = \frac{3.0 \times 10^8}{5 \times 10^{-7}} = 6 \times 10^{14} \text{ Hz}$$

As frequencies go, this is extraordinarily high; in fact, it far exceeded any frequency known at the time of Young's work. Here we have used the unit hertz (Hz) for frequency, which is equivalent to the older unit, cycles per second (cps).

18-8 COLOR AND VISION

In the interference pattern produced by *white* light through a pair of closely spaced slits, the central fringe is white, but the side fringes show color (see Fig. 18-23, inside back cover). Each color has its own set of equally spaced bright fringes, the violet fringes being the closest together and the red fringes the farthest apart. Fringes of the other colors of the rainbow are separated by distances intermediate to these two extremes. This makes sense if we associate color of light with a distinctive wavelength, a short wavelength ($\sim$4000 Å) for violet light and a long wavelength ($\sim$7000 Å) for red light. Perhaps there are similar disturbances of even shorter or longer wavelengths: all we can say now is that the average eye is not sensitive to them.

There is a curious contrast between the human ear and eye. If one takes a sound wave at the lowest frequency to which the ear responds (about 20 Hz), the frequency can be increased by a factor of 1,000 and the sound will still be heard; but if one has a light wave of the lowest perceptible frequency, about $(3 \times 10^8 \text{ m/s})/(7 \times 10^{-7} \text{ m}) = 4 \times 10^{14}$ Hz, it is not possible even to double this frequency before the light becomes invisible. The "light keyboard" extends less than 1 octave!

Acoustic and visible spectra compared

EXAMPLE 18-2

Monochromatic (single-wavelength) light passes through a narrow slit, then through two distant parallel slits separated by 0.20 mm, and finally falls upon a screen located 1.0 m beyond the plane of the parallel slits. Alternating bright and dark fringes are seen on the screen, the distance between adjacent bright fringes being 3.0 mm. What is the wavelength of the light?

SOLUTION

The arrangement is just that of the Young double-slit experiment involving small angles, and [18-9] applies here. Since

$$\frac{y_n}{D} \approx \frac{n\lambda}{d}$$

we find for $y_1 = 3.0$ mm,

$$\lambda \approx \frac{y_1 d}{D} = \frac{(0.20) \times 10^{-3} \text{ m})(3.0 \times 10^{-3} \text{ m})}{1 \text{ m}}$$
$$= 6.0 \times 10^{-7} \text{ m} = 6000 \text{ Å}$$

EXAMPLE 18-3

Two identical radio oscillators generate identical waves of wavelength λ which spread out from the antennas equally in all directions. The antennas are separated by a distance $\frac{1}{2}\lambda$ and lie along a north-south line. Consider all the points a fixed distance from the point midway between the two antennas. At which of these points is the resultant wave disturbance a maximum? At which is it a minimum?

SOLUTION

The resultant wave disturbance is a maximum at the points directly to the east and to the west of the two oscillators. The separate waves

FIGURE 18-24

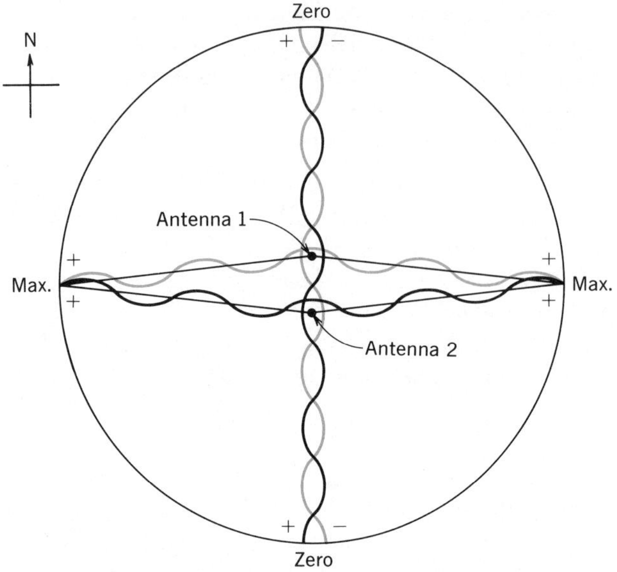

always combine constructively because the path lengths to the two oscillators are exactly the same (see Fig. 18-24).

Query Assume that the two oscillators in Example 18-3 do not oscillate together. Instead one oscillator and its antenna generate a wave crest at the same instant the other oscillator and antenna generate a wave trough. In what directions is the resultant wave disturbance now a maximum and minimum?

But the waves interfere destructively at the points directly north or directly south of the two antennas, since the two path lengths differ by exactly $\frac{1}{2}\lambda$. Unlike the usual case of the double slit, in which there are many dark fringes, there are no other directions here in which the net wave disturbance is zero. Note, here the angles are large, and [18-8] and [18-9] do not apply.

+ 18-9 THE GRATING SPECTROMETER

The grating spectrometer, a delicate instrument for measuring the wavelengths of light with very high precision, is a valuable device because much information about the internal structures of atoms and molecules can be deduced from the wavelengths of light they emit. A grating is merely a refinement of Young's apparatus, with a very large number N of narrow slits, each separated from the next by a very small distance d. One form of the so-called transmission grating for visible light consists of a large number of equally spaced parallel straight lines scratched on a plate of glass.

The grating: a collection of narrow equally spaced parallel slits

A grating of six slits (Fig. 18-25) will illustrate the main features (real instruments have thousands of slits). The spacing d is much less than D, the distance from the grating to the screen. In fact, D

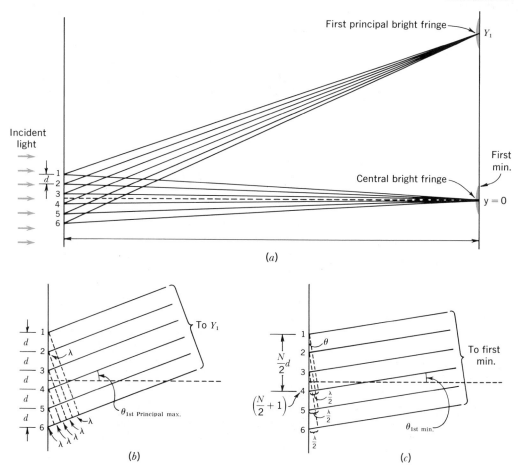

FIGURE 18-25 (*a*) The central and first principal bright fringes in a six-slit experiment. (*b*) The essential distances which determine the constructive interference at the first principal bright fringe. (*c*) The essential distances that determine the destructive interference at the first dark fringe away from the central bright fringe.

is so large compared to Nd that the distances from the *center* of the screen to the various slits differ among themselves by much less than λ. Clearly, then, at the central position $y = 0$ on the screen the waves from all six sources arrive to reinforce each other, and this position is one of brightness. Since the light comes from a large number of slits, rather than two, this maximum is far brighter than that for a pair of slits. (In fact, it is 9 times brighter, not 3 times, because the light intensity, or the energy arriving per unit area per unit time, is proportional to the *square* of the number of individual disturbances combining constructively.[1]) Another major maximum in intensity is at the point Y_1, where the distances of the sources 1 and 2 from the screen differ by exactly 1 wavelength, just as in the earlier case

[1] This is also true for a disturbance transmitted in a medium consisting of coupled springs. The potential energy of a spring is proportional to the *square* of the deformation ΔL of its relaxed length [9-3]. If three identical superimposed disturbances cause a threefold deformation of the spring, the potential energy is increased ninefold.

[1] The rays from all the slits to a point on the screen will be very nearly parallel provided the overall width of the grating $(N - 1)d$ is much less than D. If this is not the case, a converging lens can be interposed between grating and screen, with the screen in the focal plane of the lens. All rays leaving the grating in the same direction will then be brought to the same point of the screen.

of two slits. But Fig. 18-25b shows that if sources 1 and 2 differ in path length by 1 wavelength, so do sources 2 and 3; indeed, all pairs of adjacent sources interfere constructively with each other to produce a strong maximum in brightness.[1] It follows from this reasoning that all the major maxima have the same locations as in the double-slit arrangement [18-9]

[18-10] $$Y_n \simeq \frac{n\lambda D}{d} \qquad \text{where } n = 0, 1, 2, \ldots$$

The fringes from a grating are bright and narrow.

Not only does the grating produce very *bright* principal maxima, but these maxima are very much *narrower* than the fringes produced by just two slits, as can be seen from Fig. 18-25c. Here the angle θ is that corresponding to the brightness minimum nearest $y = 0$. Since the difference in distance between sources 1 and 4 and the point observed on the screen is $\frac{1}{2}\lambda$, these two sources interfere destructively. So do the sources 2 and 5, and 3 and 6, which also differ by $\frac{1}{2}\lambda$. In short, there is no net wave disturbance at this small angle, which differs only slightly from that for which all slits interfere constructively. If $d \ll D$ and the angle $\theta' \simeq \theta$ is small, then:

For N slits:

$$\theta_{\text{1st min.}} \simeq \frac{\frac{1}{2}\lambda}{\frac{1}{2}Nd}$$

[18-11] $$y_{\text{1st min.}} \simeq D\theta_{\text{1st min.}} \simeq \frac{\lambda D}{Nd}$$

Thus, in a grating the *width* of the central principal maximum (distance from the first minimum on one side to the first minimum on the other side) is $2\lambda D/Nd$. The larger the number of slits, the narrower the fringe. (The double slit is merely a grating with $N = 2$.) We have deduced this result for the central bright fringe, but a similar argument can be used to confirm the same result for all other principal fringes, at least so long as the angle θ to the fringe is small. There are weak subsidiary peaks between the principal fringes we have discussed, but they can be neglected when N is large.

It is also possible to deduce the dependence of fringe width on N from energy considerations. Although the argument is sophisticated and qualitative, it is quick and general, features which physicists like. If you increase the number of slits by a factor F (we went from 2 to 6 slits, corresponding to $F = 3$), the total energy reaching the screen increases by the factor F. But, as we have already stated, the energy *per unit area* at the center of a given fringe varies as the *square* of the number of slits. The only way the energy per unit area can

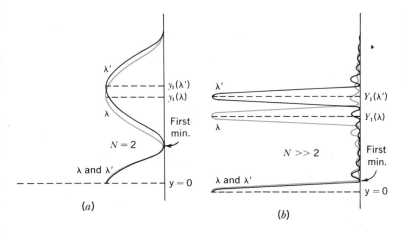

FIGURE 18-26 (a) Unresolved interference fringes in a double-slit experiment for two slightly different wavelengths. (b) Resolved interference fringes in a grating experiment for the same light of two slightly different wavelengths.

increase by a factor F^2 while the total energy increases only by a factor F is to have the fringe get narrower at the same time that it gets more intense. Indeed, the width must decrease by the factor $1/F$, which is just what was found in [18-11].

In studying atomic structure, the narrowness of the grating fringes is sometimes even more important than their brightness. It is often crucial to know whether atoms are emitting light of a single frequency or of two or more nearly identical frequencies. Discrimination is impossible if the fringes are broad (see Fig. 18-26a), but if the number of slits is large and the fringes narrow, one can distinguish between (resolve) two signals, as shown in Fig. 18-26b. Some of the detailed structure of yellow sodium light is shown in Fig. 18-27. These two distinct wavelengths near 6000 Å and differing by only one part in 1,000 are nevertheless resolved by the grating.

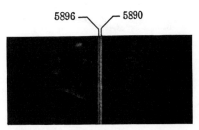

FIGURE 18-27 Sodium-light fringes obtained with a grating. The two distinct lines indicate that sodium atoms emit light having two distinct (but very nearly equal) wavelengths of 5890 Å and 5896 Å.

A many-slit grating can resolve close differences in wavelength.

White light (4000 to 7000 Å) is sent through a transmission grating with 10,000 lines per centimeter. What is the angular separation between the red and violet extremes of the spectrum for $n = 1$ (called the *first-order spectrum*)?

EXAMPLE 18-4

The slits in the grating are separated by $d = 1/10{,}000$ cm $= 10^{-6}$ m. For the red light of wavelength 7.0×10^{-7} m

SOLUTION

$$d \sin \theta \approx n\lambda$$

$$\theta \approx \sin^{-1}\frac{n\lambda}{d} = \sin^{-1}\frac{1 \times 7 \times 10^{-7}}{10^{-6}} \approx \sin^{-1} 0.7 \approx 45°$$

And for the violet light of wavelength 4.0×10^{-7} m,

$$\theta \approx \sin^{-1} 0.4 \approx 23°$$

The entire visible spectrum is thus spread over $22°$. Note that for this grating [18-8] and [18-9], which require that θ be small, cannot be used.

EXAMPLE 18-5

Light of wavelength λ passes through a grating with 1,000 slits, adjacent slits being separated by the distance d. The transmitted light strikes a screen a large distance $(D \gg 1,000d)$ from the grating. What is the intensity on the screen for light leaving the grating in such a direction that the difference in path lengths from a point on the screen to the slit at the top and to the slit at the bottom of the grating is (a) λ, (b) $\frac{1}{2}\lambda$, (c) 2λ?

SOLUTION

a   For the path difference $\Delta = \lambda$ there is *total* darkness on the screen. As Fig. 18-28 shows, for every source of light (slit) in the top half of the grating there is another source in the lower half which is exactly $\frac{1}{2}\Delta = \frac{1}{2}\lambda$ farther away. The light from slit 1 destructively interferes with light from slit 501; likewise for slits 2 and 502, 3 and 503, etc.

b   When $\Delta = \frac{1}{2}\lambda$, there is neither completely destructive interference nor completely constructive interference. To be sure, the disturbances from slits 1 and 1,000 interfere destructively, but at the same time the disturbances from slits 500 and 501 very nearly combine constructively. The only situation in which total darkness results is when *all* the slits can be matched off in pairs, with the path difference exactly $\frac{1}{2}\lambda$ for each pair.

c   If $\Delta = 2\lambda$, again there is total darkness, as solution (a) then applies separately to the upper and lower halves of the grating.

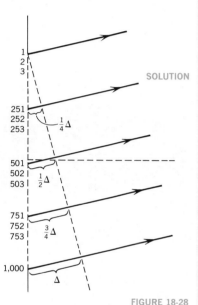

FIGURE 18-28

+   18-10   NEWTON'S RINGS

Newton's rings, a phenomenon easily explained by the interference of waves

The phenomenon of Newton's rings (Section 17-7) finds a ready explanation in terms of the interference effects between waves of light. First consider the situation in which light shines on the glass lens and plate from above and the reflected light is viewed from above, as in Fig. 18-29. Rays of light that originate from the same source and reach the eye by reflection from the curved surface of the lens and the flat surface of the lower plate of glass can produce interference effects. The two points at which the rays are reflected are separated by the distance h. Thus, the ray from the lower surface travels a distance $2h$ farther than the ray from the upper surface.

Whether the combined wave disturbance reaching the eye shows brightness or darkness depends upon the size of $2h$. If one sees brightness at some one spot, then another spot farther away from the center, for which the distance $2h$ is greater by just $\frac{1}{2}\lambda$, will show darkness. The whole pattern consists of a number of concentric rings, alternately bright and dark and centered about the point where the two pieces of glass touch (see Fig. 17-19).

One observation is more difficult to understand: the central spot is *dark*. When transmitted (rather than reflected) light is viewed, alternating rings again are seen, but in this case the central spot is *bright*, the pattern thus being the reverse of that produced by reflection. As Fig. 18-30 shows, the path differences $2h$ are everywhere identical in the two cases. Some other factor must therefore be responsible for the interchange of bright and dark. Let us recall what we have learned about mechanical waves: they experience a polarity change when they are *reflected* by a denser medium but no polarity change when reflected by a rarer medium. Moreover, they suffer no polarity change when *transmitted* in either direction across the interface between two media. Perhaps light waves experience similar effects?

Indeed, it is easy to account for the difference between the reflected and transmitted Newton's rings if we suppose that light waves, too, show no change in polarity when transmitted across an interface or when reflected from one kind (either denser or rarer) of medium but that they do change when reflected from the other kind of medium. Thus, at a location near the point of contact ($2h \ll \lambda$) in the reflection experiment, the destructive interference must be due to polarity changes *at* the surfaces since the path length is too short to account for any change. One ray has been reflected *once by a denser* medium while the other ray has been reflected *once by a rarer* medium. By our assumptions, then, one of these two rays, and only one, experiences a polarity reversal. Completely destructive interference results, and a dark central spot is observed. Note that we have not said (indeed, we cannot say on the basis of this experiment alone) which of the two rays experiences the change.[1]

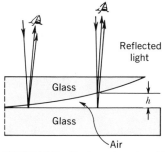

Reflected
light

FIGURE 18-29 Newton's rings viewed in reflected light arise from the interference of light reflected from the lower surface of the upper lens and from the upper surface of the lower flat plate.

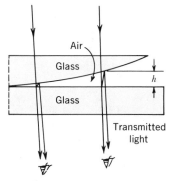

Transmitted
light

FIGURE 18-30 Newton's rings viewed in transmitted light arise from the interference of light transmitted directly through the lens and plate and light doubly reflected before being transmitted through the lower plate.

[1] It seems natural to ask whether additional reflections in the wedge of air contribute to the interference pattern. They do contribute to the magnitude of the intensity pattern, affecting the contrast between bright and dark fringes, but the position of the fringes is just as predicted on the basis of first-order reflections alone.

Consider the arrangement shown in Fig. 18-31, where two plates of glass with parallel flat faces are in contact at one end but separated at the other end by the thickness of a sheet of paper, 5.0×10^{-3} cm. The length of the glass plates is 4.0 cm. (*a*) What will you see when you view the light reflected from the two plates? (*b*) If the light has a wavelength of 6.0×10^{-5} cm, how many dark fringes will you see between the two ends of the plates?

EXAMPLE 18-6

FIGURE 18-31

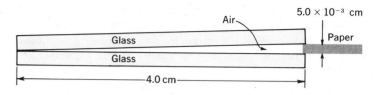

SOLUTION

Query Suppose that the curved surface of the upper glass piece in a Newton's rings arrangement is slightly irregular, with a slight protuberance or depression, perhaps only of 10^{-6} m. How can this irregularity be detected?

a The situation is like that of Newton's rings, but the air space between the two glass plates increases in thickness uniformly from left to right. Therefore, there will be alternating dark and bright straight-line fringes. Between two like fringes the difference in path length $2h$ changes by 1 wavelength. Note that, in fact, the fringes will be perfectly straight, equally spaced, and parallel only if the surfaces of the plates are themselves perfectly flat. This provides a simple but highly precise test for optical flatness, i.e., flatness of a surface to within a fraction of a wavelength of light. Any depression or protuberance is manifested in the interference pattern by bent or unequally spaced fringes.

b Since one additional dark or bright fringe is generated every time the path difference $2h$ changes by λ or the thickness h changes by $\frac{1}{2}\lambda$, the total number N of dark fringes seen with light of 6.0×10^{-5} cm wavelength as the thickness h goes from zero at one end to 5.0×10^{-3} cm at the other end is $N = (5.0 \times 10^{-3}$ cm$)/\frac{1}{2}(6.0 \times 10^{-5}$ cm$) = 166$. Conversely, counting the total number of interference fringes (or the number of fringes per unit length) and measuring the thickness of paper, one can compute the wavelength of the light.

18-11 STATUS REPORT
ON THE WAVE AND
PARTICLE THEORIES
OF LIGHT

How things stand: the wave theory accounts tor observed effects.

It is useful to review at this point how things stand with the wave and particle theories of light. As we have just seen, only the wave theory can make sense out of Young's double-slit experiment and the phenomenon of Newton's rings. These interference effects are fairly subtle because the wavelength of visible light is small compared to the size of ordinary objects.

The observation that two beams of light can cross without affecting each other (also difficult to understand on the basis of a particle theory) is at once explained by the superposition property of wave disturbances.

That the intensity of light from a point source varies inversely as the square of the distance from the source is explained by the particle theory, but it is equally well explained by the wave theory. As a wave spreads outward in all directions from a small source, the energy carried by the wave is diluted in space, so that when the distance from the point source is doubled, the energy is spread over a spherical surface which is 4 times as great.

We have seen that waves undergo reflection at the boundary between two media, so that the corresponding behavior of light is satisfactorily accounted for by the wave theory. But if this theory is to be truly acceptable it must explain *all* effects observed for light. It should:

1 Account for the rectilinear propagation of light and the relatively sharp shadows cast by sharp, opaque edges when illumined by a point source. (Section 19-1)
2 Lead to Snell's law and predict a *slower* light speed in denser media. (Section 19-3)
3 Predict a force arising from the reflection or absorption of light by objects. (Section 20-4)
4 Include a description of what the waves are waving and tell why the speed of light in vacuum is always 3.0×10^8 m/s. (Section 20-1)

In addition, we may well expect the wave theory to predict phenomena hitherto not discussed at all, and these too must check with observations.

SUMMARY

Energy (and momentum) can be propagated by a *wave* through a macroscopic distance without the transfer of mass through this distance.

A mechanical wave along a line of particles, each of mass m and coupled to the adjacent two particles by springs of length L_0 and stiffness k, moves with a velocity v that is independent of the amount of energy being propagated but is dependent on the properties of the medium (m, k, and L_0).

When two wave disturbances reach the same point simultaneously, the net disturbance is the algebraic sum (superposition) of the separate disturbances.

For a periodic wave, the wavelength λ is the distance the disturbance propagates in one period

$$\lambda = vT \qquad \qquad [18\text{-}4]$$

The alternating dark and light fringes in Young's double-slit experiment can be accounted for by the superposition of two waves, one from each slit. If the difference between the distances to the two slits is Δ, then for constructive interference (bright fringes):

[18-6] $\Delta_n = n\lambda$ where $n = 0, 1, 2, \ldots$

PROBLEMS

18-1   What happens to the wave speed of a line of identical particles interconnected by identical springs (Fig. 18-2) if (a) the mass of each particle is increased fourfold, (b) the stiffness constant of each spring is increased fourfold, and (c) the separation distance between adjacent particles is increased fourfold? (Hint: show first that the equivalent k for 4 springs connected end to end in series is $\frac{1}{4}$ of k for a single spring.)

18-2   The two pulses shown in the figure travel in opposite directions along a rope. Use the superposition principle to show the resulting waveforms at times (a) shortly after the pulses begin to collide, (b) about at the middle of the collision, and (c) well after the collision is completed.

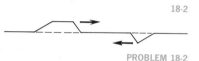

PROBLEM 18-2

18-3   The human ear is sensitive to sound frequencies in the range of about 20 to 20,000 Hz. If the speed of sound in a given gas medium is 330 m/s, what is the range of wavelengths corresponding to these frequencies?

18-4   Which of the following five quantities are uniquely determined by the source; by the medium; or by neither alone: (a) frequency, (b) wavelength, (c) speed of propagation, (d) period, and (e) amplitude?

18-5   Two long sinusoidal disturbances each having the same amplitude and frequency and each with a positive half cycle at the front end (see figure) travel in opposite directions along the same stretched rope and collide at the point x = 0. Describe qualitatively (with the help of sketches) the waveforms within a few wavelengths of x = 0 for several different times over a time interval of several periods well after the disturbances first collide.

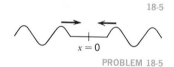

x = 0

PROBLEM 18-5

18-6   A long sinusoidal disturbance (like the one initially on the left in Problem 18-5) travels along a stretched rope securely anchored to a massive wall at its right end (see Fig. 18-13). The disturbance is reflected by the wall. Describe qualitatively the resulting waveform by drawing several sketches showing the portion of the rope within several wavelengths of the wall at several different times over a time interval of several periods well after the disturbance first hits the wall.

18-7   Light of wavelength 5.0×10^{-7} m is incident normally on an opaque barrier containing two narrow parallel slits. The distance between adjacent bright fringes near the center of a screen 1.2 m beyond the plane of the

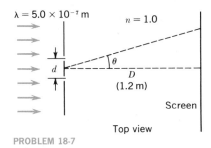

PROBLEM 18-7

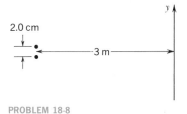

PROBLEM 18-8

slits is 0.4 cm. The space between the plane of the slits and the parallel plane of the screen is evacuated. (a) What is the distance d between the slits? (b) List four independent changes in the size of quantities identified in the figure, each one of which by itself would reduce the separation between adjacent fringes to *half* of that given above.

Two *identical parallel line sources* emit waves uniformly in all directions perpendicular to the line of the source. The sources are 2.0 cm apart. A screen parallel to the plane of the sources is placed 3.0 m away, as shown in the figure. Find the positions on the screen nearest to the center (y = 0) where the average energy received is *minimum* if (a) the wavelength is 2.0 cm, (b) the wavelength is 2.0×10^{-5} cm, (c) the wavelength is 2.0 cm, but the two otherwise identical sources emit disturbances of equal magnitude but opposite sign, (d) the wavelength is 20.0 cm, but the sources are again completely identical as in (a) and (b).

18-8

Two transmitters A and B emit identical signals which are received in the plane CC' shown in the figure. There is an intensity maximum at the central position y = 0. The nearest intensity minima are at the positions y = ±35 cm. (a) What is the wavelength of the signals? (b) What would you know about the signals emitted at A and B if a minimum had been observed at y = 0 and maxima at y = ±35 cm?

18-9

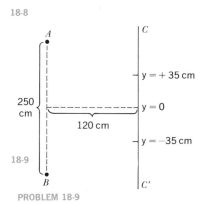

PROBLEM 18-9

Two very thin lamp filament wires are placed parallel to each other a distance 0.1 mm apart. When the two filaments are heated to the brightness of an ordinary incandescent bulb, do you see interference fringes on a distant screen parallel to the plane of the filaments? Explain.

18-10

Two radio oscillators generate *identical* signals of frequency 30×10^6 Hz (30 MHz). The identical radio waves spread out from each antenna equally in all directions with speed 3×10^8 m/s. At points on a circle of radius 3×10^4 m centered at a point midway between the two antenna maximum signals are received on the north-south and east-west lines only, and minimum signal is received on lines making 30° angles with the east-west line. (a) What is the distance in meters between the two antennas? (b) What is the orientation of the line joining the two antennas? Explain your reasoning.

18-11

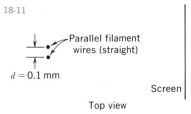

PROBLEM 18-10

18-12   A thin wedge of air is formed by slipping a thin piece of paper between the right-hand edges of two glass plates, as in Fig. 18-31. The left-hand edges of the plates remain in contact with each other. If normally incident light of wavelength 6×10^{-7} m exhibits 40 interference fringes in the 5.0-cm distance between the edges of the glass plates, how thick is the paper?

18-13   Consider the light-intensity pattern on a screen placed a large distance beyond a grating. How do the following changes affect the positions of the principal bright fringes: (a) doubling the slit spacing, (b) doubling the number of slits, (c) doubling the wavelength of the incident light?

18-14   A grating 1.0 cm wide consists of 1,000 narrow parallel equally spaced slits in an otherwise opaque screen. A beam of light from a small distant line source is transmitted through the grating. This light, which originates from a sodium lamp, consists of two wavelengths, $\lambda_1 = 5.890 \times 10^{-7}$ m and $\lambda_2 = 5.896 \times 10^{-7}$ m. Fringes are observed on a distant screen. Assuming that all angles are small, find the approximate angular positions θ and the approximate angular widths $\Delta\theta$ (the angle subtended at the grating by the lines drawn to the first two minima on either side of a fringe) for the first bright fringes away from the center of the screen for (a) λ_1 with all slits covered except two center ones, (b) λ_2 with all slits covered except two center ones, (c) λ_1 with all slits uncovered, (d) λ_2 with all slits uncovered. (e) Can you always tell from the pattern on the screen that there are two separate wavelengths in this light?

PROBLEM 18-11

18-15   Show that any function $f(x - vt)$ represents a wave *traveling* in the positive x direction while $f(x + vt)$ represents a wave traveling in the negative x direction. *Hint:* Consider the shape and location of some disturbance $f(x - vt)$ at time $t_0 + \Delta t$?

18-16   (a) Show that a traveling wave having a sinusoidal shape and moving in the positive x direction can be represented by $\sin(kx - \omega t)$, where $k = \omega/v$ and k is a constant which converts distances into angles (measured in radians). We choose the origin $x = 0$ such that at $t = 0$ the disturbance is zero at $x = 0$. (b) Show that $\omega = 2\pi/T$, where T is the period of the disturbance, and $k = 2\pi/\lambda$, where λ is the wavelength.

DIFFRACTION

CHAPTER
NINETEEN

This chapter consolidates the position of the wave theory by showing that not only the spreading of light around corners but even its familiar straight-line propagation in a beam can be explained as an interference phenomenon. Here we shall be involved with the superposition of waves from a continuous distribution of sources, a type of interference known as *diffraction*.

19-1 PROPAGATION OF WAVES AS A BEAM

A direct demonstration of waves traveling as a beam can be made with periodic waves on the surface of water. When such waves impinge upon an absorbing barrier having a very narrow opening or window (Fig. 19-1), the transmitted wave is a series of concentric semicircles, of alternately positive and negative disturbance, which spreads out into the entire region behind the barrier; however, when the window is wide (Fig. 19-2), parallel lines of positive uniform disturbance alternate with similar parallel lines of negative disturbance in the region directly beyond the opening, but essentially no disturbance appears anywhere within the geometrical shadow of the barrier. The transmitted wave is thus a beam, or a bundle of rays, just as wide as the window and traveling perpendicular to it. To be sure, at sufficiently large distances beyond the barrier the disturbance fans out again, but there will be a well-defined beam at all

FIGURE 19-1 (*a*) Water waves incident upon a barrier with a very narrow (less than 1 wavelength) opening. The transmitted waves form circular patterns. (*Film Studio, Education Development Center.*) (*b*) Sketch of the advancing disturbances in (*a*) at an instant when there is maximum positive disturbance at the opening.

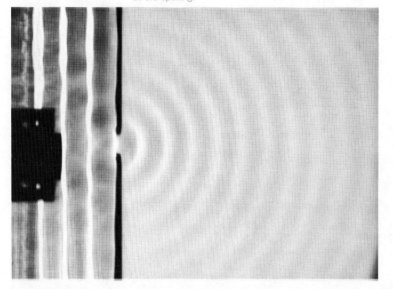

(*a*)

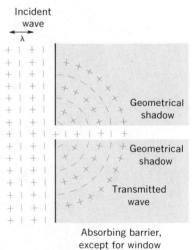

Incident
wave

Geometrical
shadow

Geometrical
shadow

Transmitted
wave

Absorbing barrier,
except for window

(*b*)

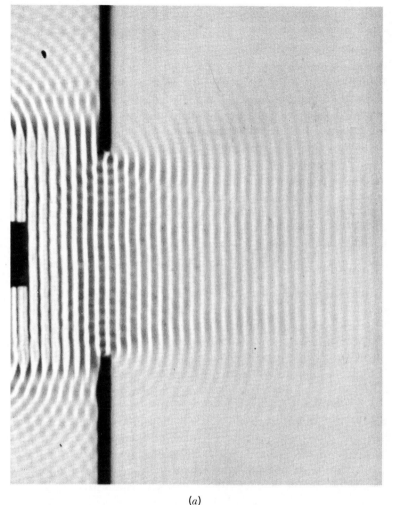

(a)

Definition of Δ^*: $\Delta^* = D_E - D_0$. Here D_0 shown in the figure below is the distance from the center of the screen to the center of the slit and D_E is the distance from the center of the screen to the edge of the slit. If $\Delta^* \ll D_0$, then $\Delta^* \cong w^2/8D_0$ (using an approximation similar to that in Example 19-2).

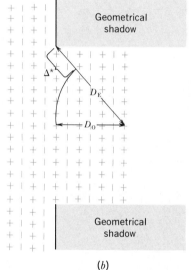

(b)

FIGURE 19-2 (a) Water waves incident upon a barrier with a wide (more than several wavelengths) opening. The transmitted waves have the essential characteristic of a beam: the width of the transmitted beam disturbance is approximately equal to the width of the opening. (*Film Studio, Education Development Center.*) (b) Sketch of the advancing disturbance in (a) at an instant when there is maximum positive disturbance at the opening.

Diffraction: the superposition of waves from infinitely many adjacent sources

distances directly beyond the window for which the lengths D_0 and D_E defined in Fig. 19-2b satisfy, as we shall now prove,

$$\Delta^* = D_E - D_0 \gg \lambda \qquad [19\text{-}1]$$

Notice that according to [19-1] there will be no well-defined beam unless the opening is many wavelengths in width.

Wherever condition [19-1] is satisfied, we obtain the beam as a *diffraction pattern*, the superposition of waves from *infinitely many adjacent* sources, in this case distributed over the opening in the barrier. Hard though it may be to believe that a straight beam is an interference effect, the type of reasoning which so successfully accounted for Young's experiment works here too.

After assuming that the wave sources in the opening are all identical in intensity and in the variation of their periodic signals with time, we consider groups, or zones, of these oscillators such that the disturbance from one zone will annul, or nearly annul, that from a neighboring zone. This idea goes back to the French physicist Augustin Fresnel (1788–1827), who introduced it around 1815.

Figure 19-3 shows how the zones, labeled 1, 2, 3, . . . and 1', 2', 3', . . . , are chosen for a given observation point P. The path length to P from the lowest source in any one zone differs from the path length to P from the highest source in the same zone by exactly $\frac{1}{2}\lambda$. For example, the distance b exceeds a by $\frac{1}{2}\lambda$, and c also exceeds b by $\frac{1}{2}\lambda$. Nearly every source in zone 1 can be matched off against a corresponding source in zone 2 such that the path length for the source in zone 2 exceeds that for the source in zone 1 by $\frac{1}{2}\lambda$. If we represent the net disturbance from zone 1 by an upward vector, the net disturbance from zone 2 is a downward vector of nearly the same length, as in Fig. 19-4. Indeed, the sense of the disturbance alternates from zone to zone, and the disturbance from one zone tends to cancel that from its neighbor. The cancellation is not complete, however, because the magnitude of the disturbance diminishes slightly for zones farther off the axis. The main reason for this decrease is simply that the higher-numbered zones are progressively smaller (see Fig. 19-3) and hence have progressively smaller total outputs. In addition, they are somewhat farther away from point P, so that their energy is more dispersed by the time it reaches P. (As we shall see later, there is still a third reason why the more distant zones are less important than the nearby zones: the oscillators will be found to emit energy more strongly in the forward direction than in oblique directions.) Consequently, the contributions of all zones, primed and unprimed, always give a nonzero resultant amplitude for point P at the central position.

Now suppose that the point P is shifted in the plane of observation away from the central position but not so far up or down that it is near the geometrical shadow of the barrier (Fig. 19-5b). Then the condition given by [19-1] simply guarantees that there still is a large number of zones *both* above and below point P, and the amplitude of the resultant wave disturbance will be the same as when P was at the central position. This is easy to understand, since the resultant disturbance is determined primarily by the important nearest zones,

The key to treating diffraction: consider groups of oscillators, or zones

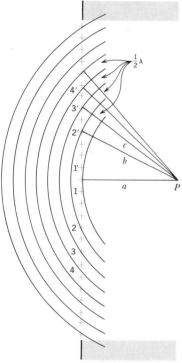

FIGURE 19-3 Fresnel's method for constructing zones within the opening shown in Fig. 19-2. Starting at the central point P in the plane of observation, one imagines a series of circular arcs, the first with a radius a equal to the distance from point P to the opening and the others increasing in radius by incremental steps of $\frac{1}{2}$ wavelength. If each zone is considered as a source of disturbances similar to the narrow opening of Fig. 19-1, the disturbances from adjacent zones will interfere destructively with each other while those from alternate zones will interfere constructively.

with a negligible contribution from the distant zones. In the new position for point P there are still as many nearest zones as for the central position; only the arrangement of the distant zones has changed. Thus the important contribution from the near zones is the same as before.

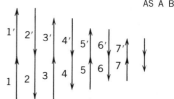

FIGURE 19-4   The addition (superposition) of the disturbances at point P arising from the various zones of Fig. 19-3. The two central zones (1 and 1′) produce identical disturbances at point P and add constructively. Then the polarities of the successive disturbances alternate with zone number, and the magnitudes slowly decrease with increasing number, leaving a net light intensity at P.

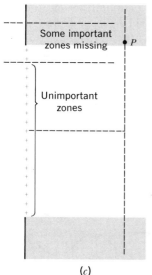

(a)

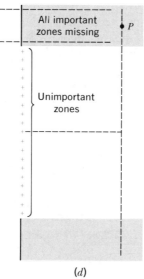

(b)

(c)

(d)

FIGURE 19-5   The beam transmitted by the wide opening of Fig. 19-2 (a) The net disturbance at point P directly in front of the center of the opening has the value obtained in Fig. 19-4 and is predominantly determined by a limited number of zones near the center of the opening. (b) If the point P is shifted from the center but kept far enough from the geometrical shadow so that the new set of zones still includes as many important zones as in (a), the net disturbance will be the same as at the center. (c) However, if the point P is moved so close to the geometrical shadow that some of these important zones are cut off by the barrier, the net disturbance will be different than at the center. (d) Indeed, if the point P is moved far into the geometrical shadow, all the important zones will be cut off by the barrier, and there will be complete darkness. The edge of the beam near the geometrical shadow will not be sharp; there is some fluctuation of the light intensity just outside the geometrical shadow, and there is some residual light intensity within this shadow (see Fig. 19-9.)

Of course, the above conclusion is applicable only if point P is far from the geometrical shadow. If P is near enough to the shadow (Fig. 19-5c), some of the important nearby zones are nonexistent; there are no oscillators in the barrier itself, only in the opening. Indeed, if P is well within the geometrical shadow (Fig. 19-5d), there are no nearby zones whatever and the net displacement for all practical purposes is zero.

Around the edge of the geometrical shadow there are some peculiar variations in the amplitude of the displacement, but this region of variation is usually small, particularly if the wavelength is small, as with visible light. Thus, for light especially, we can say that a uniform rectilinear *beam* is well accounted for by wave diffraction.

The propagation of a beam is accounted for.

19-2 WAVEFRONTS AND HUYGENS' PRINCIPLE

In Figs. 19-1 and 19-2 it is easy to identify lines (surfaces in three dimensions) along which adjacent points simultaneously experience their maximum (crests) or minimum (troughs) disturbance. Similarly we can identify lines (surfaces) which are the locus of adjacent points simultaneously experiencing *any* constant fraction of their maximum disturbance. Such a line (surface) is called a *wavefront*. In Fig. 19-1 the wavefronts are curved, but in Fig. 19-2 they are straight, at least directly in front of and near the opening.

Wavefront defined

Huygens' principle: a geometrical procedure for finding a future wavefront from a present wavefront

Wavefronts advance with the speed of the wave. After a short time Δt the wavefront (crest) through P' in Fig. 19-6 will have advanced to a line through P, and the wavefront (trough) which was at P at the beginning of this time interval will have advanced to P''. A simple geometrical procedure, known as *Huygens' principle,* locates any type of wavefront after a time Δt if its position at the beginning of this interval is known. Draw semicircles of radius $v \Delta t$ centered about each point on the initial wavefront; the envelope of such Huygens *wavelets* is the wavefront after the time Δt has elapsed. The con-

FIGURE 19-6 Wavefronts (surfaces in three dimensions) along which adjacent points simultaneously experience their maximum (crest) or minimum (trough) disturbance moving to the right with the same speed as the waves.

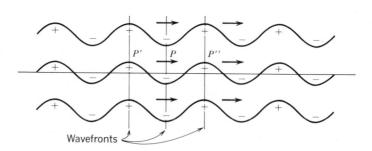

Wavefronts

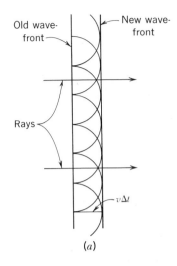

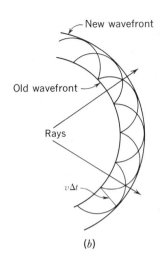

struction of both spherical and plane wavefronts is shown in Fig. 19-7. Because the procedure is valid for a plane wavefront, it is applicable to any wavefront, which can be constructed out of a sufficiently large number of small plane wavefronts.

In the particle theory of light the term *ray* meant the path followed by a particle. We still use this term in the wave theory, with the understanding that it now means the direction in which energy is transported by the advancing wavefronts. Therefore, a ray must always be a line drawn at right angles to the wavefronts.

Wave theory not only accounts for the existence of straight beams of parallel light rays but yields the correct laws for their reflection and refraction. To show this we represent a parallel beam by its plane wavefronts and make use of Huygens' principle.

19-3 REFLECTION AND REFRACTION OF WAVES

In Fig. 19-8 plane wavefront 1 approaches the interface between two media. Subsequent positions of this front, labeled 2, 3, 4, have been successively constructed by Huygens' principle. After first meeting the interface in *A*, the front is split between the two media. We separately consider the portions of the front which stay in the first medium, the *reflected wave*, (Fig. 19-8*a*) and the portions which are transmitted into the second medium, the *refracted wave* (Fig. 19-8*b*).

Law of reflection derived from Huygens' principle

To obtain the law of reflection consider (Fig. 19-8*a*) point *B* on front 2, which just reaches the interface in the time Δt used in the Huygens

FIGURE 19-8 The use of Huygens' principle to construct (*a*) a reflected beam and (*b*) a refracted beam when a beam of light is incident at angle θ_1 upon the interface between two media. The wave speed is smaller in medium 2 than in medium 1. and the angles of reflection and refraction are θ_1' and θ_2, respectively.

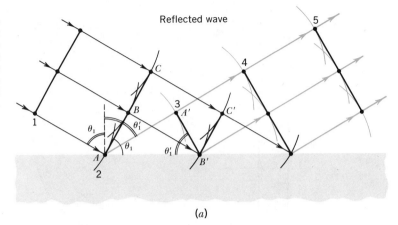

(*a*)

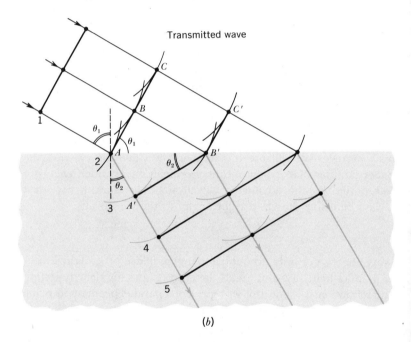

(*b*)

construction. The part of the new wavefront 3 that belongs to the reflected wave is the envelope of all the Huygens wavelets of radius $v_1 \Delta t$ emanating from section AB of wavefront 2. Here v_1 is the wave speed in medium 1. Thus the reflected front must pass through B' and must at A' be tangent to the wavelet emitted from A. This makes $AA'B'$ a right triangle, which (because $\overline{AA'} = \overline{BB'} = v_1 \Delta t$ and side AB' is common to both triangles) is congruent to the right triangle

$B'BA$. There follows at once

$$\theta_1 = \theta_1' \qquad\qquad [19\text{-}2]$$

which is the same as the relation given by particle theory.

Turning now to transmitted part of wavefront 3, we see that the same reasoning applies: $AA'B$ is a right triangle. Here, $\overline{AA'} = v_2 \Delta t$, since the wavelet has traveled through medium 2. As the two right triangles $AA'B$ and ABB' have a common hypotenuse AB',

Snell's law derived from Huygens' principle

$$\frac{\sin \theta_1}{\sin \theta_2} = \frac{\overline{BB'}}{\overline{AA'}} = \frac{v_1 \Delta t}{v_2 \Delta t} = \frac{v_1}{v_2} \qquad\qquad [19\text{-}3]$$

This is again Snell's law [17-2]: the ratio of the sines is a constant. But there is a crucial difference. Wave theory [19-3] gives v_1/v_2 as the value of the constant ratio, whereas Newtonian particle theory in [17-5] gave v_2/v_1! Recall that when a wave enters an optically denser medium, $\theta_2 < \theta_1$. Thus, wave theory predicts a *smaller* speed in the denser medium, in direct opposition to the Newtonian theory but exactly as demanded by Foucault's measurements (Section 17-6).

Query In deriving the laws of reflection and refraction we tacitly assumed that the two pieces $A'B'$ and $B'C'$ of wavefront 3 were straight lines (really planes). Can you justify this?

Comparing [19-3] with the definition [17-2] of the relative index of refraction, we see that

$$\frac{v_1}{v_2} = n_{21} \qquad\qquad [19\text{-}4]$$

By use of the result of Example 17-6, this can also be written

$$\frac{v_1}{v_2} = \frac{n_2}{n_1} \qquad\qquad [19\text{-}5]$$

where n_1 and n_2 are indices of refraction for media 1 and 2 relative to vacuum. Moreover, if we let medium 1 be vacuum and write $v_1 = c$, $v_2 = v$, $n_1 = 1$, $n_2 = n$, [19-5] becomes

$$v = \frac{c}{n} \qquad\qquad [19\text{-}6]$$

The speed of light in any medium is therefore the speed c of light through vacuum divided by the medium's refractive index.

Although the speed of light changes from one medium to the next, the frequency f (or the period $T = 1/f$) stays fixed, being determined solely by the original source. One color of light, i.e., one frequency, is the same whatever the medium through which it is transmitted. What changes along with the speed is the wavelength λ. Since $v = \lambda/T = \lambda f$, we have from [19-6]

FIGURE 19-9 Light-intensity pattern near the edge of a light beam. (*Cagnet, M., M. Françon and J. C. Thrierr, "Atlas optischer Erscheinungen," Springer-Verlag, Berlin-Göttingen-Heidelberg, 1962.*)

$$[19\text{-}7] \quad \lambda_{\text{medium}} = \frac{\lambda_{\text{vac}}}{n}$$

The wavelength is shrunk in a refractive medium by the factor n. For example, red light with a wavelength of 6000 Å in vacuum has

437

SECTION 19-4
FRESNEL AND
FRAUNHOFER
DIFFRACTION AT
A SINGLE SLIT

a wavelength in glass ($n = 1.5$) of $\lambda_{\text{medium}} = 4000$ Å. Even though there is a drastic change in wavelength, the frequency-sensitive eye still perceives the disturbance as red.

EXAMPLE 19-1

A light wave travels through a medium having an index of refraction $n = 1.20$ relative to vacuum. What is the period of this wave if its observed wavelength in this medium is 5.0×10^{-7} m?

The period is related to the wavelength by $\lambda = vT$, where the velocity $v = c/n$. Therefore

$$T = \frac{\lambda}{v} = n\frac{\lambda}{c} = 1.2\frac{5.0 \times 10^{-7}}{3.0 \times 10^8} = 2.0 \times 10^{-15} \text{ s}$$

19-4 FRESNEL AND FRAUNHOFER DIFFRACTION AT A SINGLE SLIT

The interference pattern of a single window or slit was shown to be a uniform beam, provided the condition $\Delta^* \gg \lambda$ held. We ought to say a *nearly* uniform beam because we neglected a narrow region near the border of the geometrical shadow in which the intensity fluctuates, producing fringes (see Fig. 19-9). For light, the effect is subtle, but it was noticed by the Italian physicist Francesco Maria Grimaldi (1618–1663). The name *diffraction* literally means a breaking in pieces.

Fresnel's zone analysis also furnishes the explanation of these fringes. Returning to Fig. 19-5c, for P close to the shadow line, the sequence of important nearby zones is broken off because of the presence of the barrier edge. The net wave disturbance (shown in Fig. 19-10) is then strongly influenced by whether the last zone at the barrier edge contributes a vector up or down. Consequently, as P crosses over into the geometrical shadow, the intensity will go through maxima and minima.

So much for the *beam diffraction pattern* corresponding to $\Delta^* \gg \lambda$, that is, to the slit having a great number of zones for points well within the geometrical shadow. If the slit has only a few such zones ($\Delta^* \approx \lambda$) or a small fraction of a single zone ($\Delta^* \ll \lambda$), the result is, respectively, a *Fresnel diffraction pattern* or a *Fraunhofer diffraction pattern*.[1]

Let us briefly examine these two patterns. Remember that although different names are attached to them, one and the same physical mechanism underlies them both, as well as the beam. Indeed, for a given wavelength and slit width, a beam is registered when the

[1] Joseph von Fraunhofer (1787–1826) was a German optician and physicist.

FIGURE 19-10 ᐟ(a) Fresnel zones used to obtain the net disturbance at point P near the geometrical shadow from the opening in Fig. 19-2. Because of the barrier some important zones are missing. (b) The addition (superposition) of the individual disturbances from the various zones in order to obtain the net disturbance. A comparison with Fig. 19-4 shows the effect of the missing zones on the light pattern.

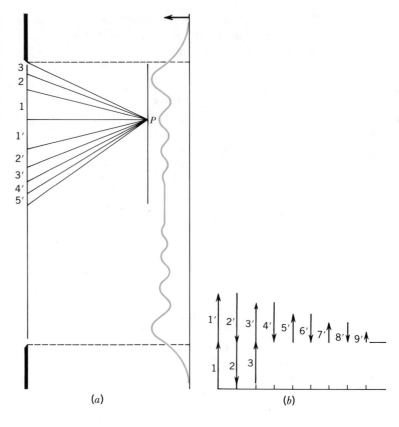

(a)

(b)

FIGURE 19-11 The light-intensity pattern (b) for the special case (a) where the opening in the barrier corresponds to exactly four Fresnel zones, two above and two below the midpoint.

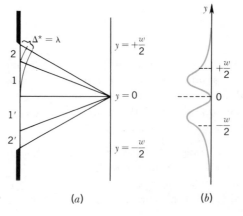

(a)

(b)

439

SECTION 19-4
FRESNEL AND
FRAUNHOFER
DIFFRACTION AT
A SINGLE SLIT

observing screen is close to the slit, a Fresnel pattern when it is farther away, and a Fraunhofer pattern when it is farther still.

Fresnel patterns can be represented by the particular case $\Delta^* = \lambda$ Fresnel diffraction (Fig. 19-11b). This pattern is typical in that its width is roughly equal to that of the slits, but it has a special feature which is even more surprising than Young's experiment. The net disturbance shows a large minimum directly behind the center of the slit. This may seem absurd, even from the standpoint of wave theory, but the explanation is obvious from Fig. 19-11a. At this center position the disturbances from the two outer zones cancel those from the two central zones. Away from the central dark line the intensity first rises to a maximum then falls to zero. Other Fresnel diffraction patterns for other values of Δ^* will show a number of diminishing peaks.

Plane sound waves of 5 cm wavelength (frequency about 6,700 Hz) are incident normally upon a tall open window 1.0 m wide. How far beyond the window should an ear be to detect a large minimum in the sound of this wavelength?

EXAMPLE 19-2

We first need a relation between the slit width w, observation distance D_0, and wavelength λ for the conditions shown in Fig. 19-11a. The distances are as shown in Fig. 19-12. Applying the Pythagorean theorem to this triangle, we have

SOLUTION

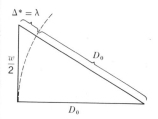

$$\left(\frac{w}{2}\right)^2 + D_0{}^2 = (D_0 + \lambda)^2 = D_0{}^2 + 2D_0\lambda + \lambda^2$$

Compared to the other terms, the term λ^2 is negligibly small; therefore,

FIGURE 19-12

$$\frac{w^2}{4} = 2D_0\lambda$$

$$D_0 = \frac{w^2}{8\lambda} = \frac{1.0^2}{8(5 \times 10^{-2})} = 2.5 \text{ m}$$

Plane light waves of 5.0×10^{-7} m wavelength (frequency about 6×10^{14} Hz) are incident normally upon a narrow slit 0.40 mm wide. How far from the plane of the slit should a screen be placed so that there is a minimum in the light at the central position?

EXAMPLE 19-3

Again, as in Example 19-2,

SOLUTION

$$D_0 = \frac{w^2}{8\lambda} = \frac{(4 \times 10^{-4})^2}{8(5 \times 10^{-7})} = 0.040 \text{ m} = 4.0 \text{ cm}$$

Compared to Example 19-2, the decrease in λ by a factor of 10^5 (this change alone would require a value of D_0, which is 10^5 times *larger* in order to keep $\Delta^* = \lambda$) is more than compensated for by the decrease in w^2 by a factor of $(4 \times 10^{-4})^2$.

Fraunhofer diffraction

A Fraunhofer diffraction pattern is characterized by wide fringes (unlike the beam pattern) and by the absence of any direct indication of the width of the slit (unlike both the beam pattern and the Fresnel pattern). See Fig. 19-13. The central fringe, in fact, may be many times wider than the slit. What is the explanation?

First of all, at *every* point on the screen directly beyond the slit there is nearly perfect constructive interference between the signals from the various identical sources in the opening. Since any differences in path length are very small relative to the critical distance of 1 wavelength, the light intensity is uniform over a width w and stays uniform until the observation point moves so far into the geometrical

FIGURE 19-13 The light-intensity patterns for viewing screens placed at three different distances beyond the barrier opening: (*a*) D_I very small; (*b*) D_{II} the same as for the special case in Fig. 19-11; (*c*) D_{III} very large. The three patterns are typical of beam propagation, Fresnel diffraction, and Fraunhofer diffraction, respectively.

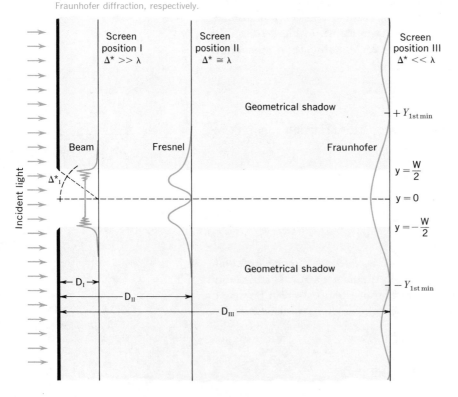

441

SECTION 19-4
FRESNEL AND
FRAUNHOFER
DIFFRACTION AT
A SINGLE SLIT

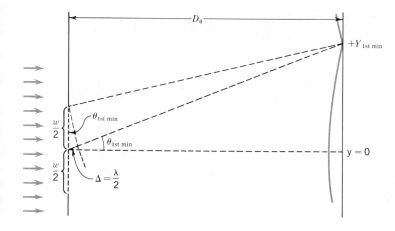

FIGURE 19-14 The essential distances used to analyze the first minimum in intensity away from the central maximum of a Fraunhofer diffraction pattern. A disturbance from any point in the lower half of the opening travels exactly $\frac{1}{2}$ wavelength farther than that from a corresponding point in the upper half; thus these two, as well as all other pairs of corresponding disturbances, interfere completely destructively.

shadow that the difference in path becomes comparable to λ. Unless the slit is less than $\frac{1}{2}$ wavelength across, which is never the case for visible light, this condition will sooner or later be realized. The first minimum in intensity will occur at that location $Y_{1st\,min}$ for which each point source in the upper half of the slit is $\lambda/2$ closer than its counterpart in the lower half of the slit.[1] The situation is pictured in Fig. 19-14, from which it is seen that

$$\sin \theta_{1st\,min} \approx \frac{\lambda/2}{w/2},$$

provided that $D_0 \gg w$. If further, $Y_{1st\,min} \ll D_0$, that is, if $\theta_{1st\,min}$ is small

$$\sin \theta_{1st\,min} \approx \theta_{1st\,min} \approx \frac{Y_{1st\,min}}{D_0}$$

Eliminating θ from these equations yields

$$Y_{1st\,min} = \frac{D_0\lambda}{w} \qquad [19\text{-}8]$$

The central fringe of the Fraunhofer pattern is therefore of width $2Y_{1st\,min} = 2D_0\lambda/w$. The interesting thing here is that w occurs in the denominator; the intensity pattern on the screen *expands* as the slit width is reduced. There is now no question of deducing the slit width from the appearance of the intensity pattern. The clear profile of the slit predicted by particle theory (and realized when $\Delta^* \gg \lambda$) will never be attained in Fraunhofer diffraction.

[1] In other words, even though the entire slit consists of but a fraction of one Fresnel zone with respect to points directly beyond it (that is the meaning of $\Delta^* \ll \lambda$), the slit is made up of two Fresnel zones for a sufficiently oblique location $Y_{1st\,min}$.

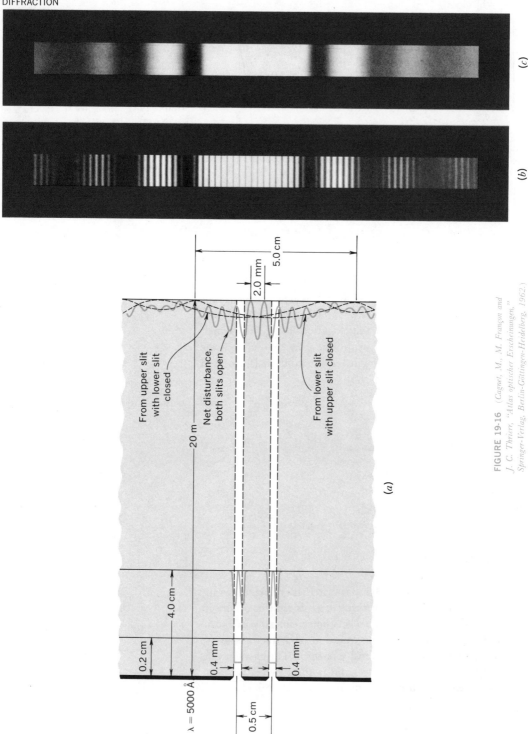

(c)

(b)

From upper slit
with lower slit
closed

Net disturbance,
both slits open

From lower slit
with upper slit closed

20 m

2.0 mm

5.0 cm

4.0 cm

0.2 cm

λ = 5000 Å

0.4 mm

0.4 mm

0.5 cm

(a)

FIGURE 19-16 *(Cagnet, M., M. Françon and J. C. Thierr, "Atlas optischer Erscheinungen," Springer-Verlag, Berlin-Göttingen-Heidelberg, 1962.)*

443

SECTION 19-4
FRESNEL AND
FRAUNHOFER
DIFFRACTION AT
A SINGLE SLIT

EXAMPLE 19-4

When a screen is placed 1.0 m from a single slit of width 0.40 mm, the diffraction pattern (Fig. 19-15) shows a central maximum bordered by two dark fringes separated by 2.50 mm. What is the wavelength of the light through the slit?

From [19-8]

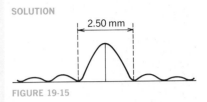

FIGURE 19-15

$$Y_{\text{1st min}} = \frac{D_0\lambda}{w}$$

or

$$\lambda = w\frac{(Y_{\text{1st min}})}{D_0} = \frac{(0.40 \times 10^{-3}\ \text{m})(1.25 \times 10^{-3}\ \text{m})}{1.0\ \text{m}}$$

$$= 5.0 \times 10^{-7}\ \text{m} = 5{,}000\ \text{Å}$$

EXAMPLE 19-5

A barrier having *two* parallel slits 0.5 cm apart and each 0.40 mm wide is illuminated by a distant narrow line source of light of wavelength 5000 Å. What light intensity pattern is observed on a screen placed (*a*) 0.2 cm beyond the barrier, (*b*) 4.0 cm beyond the barrier, (*c*) 20 m beyond the barrier?

a See Fig. 19-16*a*. At a distance of only 0.2 cm from the barrier, Δ^* is much larger than λ. As a result, there are two nonoverlapping *beams* of light from the two slits. (Δ^* defined on page 429.)

b At a distance of 4.0 cm beyond the barrier (see *Example 19-3*) the intensity patterns are still separated by 0.5 cm. Each has approximately the width of its own slit (0.2 cm), with a dark spot at the center. This is the special *Fresnel* diffraction pattern for $\Delta^* = \lambda$.

c At the other extreme, when the observation screen is placed 20 m back of the barrier, Δ^* becomes much smaller than λ. As a result, the *Fraunhofer* intensity pattern from each slit alone is very wide. We can compute the *width of the central fringe* using [19-8]

$$2Y_{\text{1st min}} = \frac{2D_0\lambda}{w} = \frac{2 \times 20\ \text{m} \times 5 \times 10^{-7}\ \text{m}}{4 \times 10^{-4}\ \text{m}} = 5.0\ \text{cm}$$

In other words, the diffraction patterns of the individual slits (shown dashed in Fig. 19-16*a*) are much wider than the separation of the slits themselves. Interference between the slits therefore becomes important. Indeed, this is simply Young's double-slit experiment, and we are now taking into account the fact that actual slits are not infinitely narrow. We use [18-6,7,8] to determine the positions of the net interference maxima, remembering now that d is the slit separation

$$y_n = n\frac{\lambda D_0}{d} = n\frac{5 \times 10^{-7} \text{ m} \times 20 \text{ m}}{5 \times 10^{-3} \text{ m}} = n(0.2 \text{ cm})$$

Thus, there will be an intensity maximum every 0.2 cm with minima midway between. This intensity pattern is shown as a solid curve in Fig. 19-16a and by a photograph in Fig. 19-16b.

A dashed curve at $D = 20$ m in Fig. 19-16a, or the photograph in Fig. 19-16c, shows the diffraction pattern observed with only the upper slit open or with only the lower slit open. The solid curve in Fig. 19-16a or the photograph in Fig. 19-16b can be described as a double-slit interference pattern (determined by the slit separation) within an envelope which is the single-slit diffraction pattern (determined by the slit width).

SUMMARY   Not only does the wave theory of light account for both the law of reflection [17-1] and the law of refraction [17-2], but it predicts that the index of refraction is

[19-4] $n_{21} = \dfrac{v_1}{v_2}$

Thus, in contrast to Newton's particle theory of light, the wave theory correctly predicts that the speed of light is smaller in the optically denser (larger n) material.

When a light wave crosses the surface from vacuum into a transparent material having an index of refraction n, the frequency (and period) of the wave remains unchanged but the velocity decreases by the factor $1/n$. Therefore, the wavelength also decreases by the same factor.

In *beam* propagation ($\Delta^* \gg \lambda$ in Fig. 19-13), the light intensity across the beam is uniform (except for fluctuations in negligibly small regions near the edges), and the width of the beam is a measure of the width of the opening. In *Fresnel* diffraction ($\Delta^* \approx \lambda$) light intensity shows major variations all across the width of the beam, but this width remains an approximate measure of the width of the opening. In *Fraunhofer* diffraction ($\Delta^* \ll \lambda$) the light intensity shows major variations all across the width of the pattern, and this width is no longer a measure of the width of the opening. Indeed, the smaller the opening, the larger the pattern on the screen.

Visible light of wavelength 5.0×10^{-7} m is incident normally on an opaque 19-1
barrier with a 2.0-m opening. The light-intensity pattern on a screen placed
parallel to the plane of the barrier is shown in the figure. What is the
distance D_0?

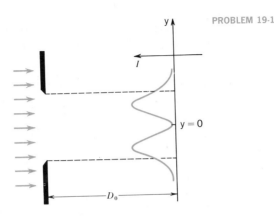

PROBLEM 19-1

Repeat Problem 19-1 for a slit of width 0.2 mm. 19-2

Repeat Problem 19-1 for a radar beam having a microwave wavelength 19-3
of 10 cm.

What is the wavelength *in water* of light having a wavelength 5.0×10^{-7} m 19-4
in vacuum. Give your answer in angstroms.

What frequency of light has a wavelength 6.0×10^{-7} m in a medium whose 19-5
index of refraction relative to vacuum is 1.5?

In the figure you are shown light-intensity patterns from a single slit. The 19-6
distance between the observation screen and the slit is kept fixed, but the
slit width w and the wavelength λ are variable. What happens to the width
of the top pattern (*a*) when λ is increased slightly (w fixed)? (*b*) when w
is increased slightly (λ fixed)? What happens to the width of the bottom
pattern (*c*) when λ is increased slightly (w fixed)? (*d*) when w is increased
slightly (λ fixed)?

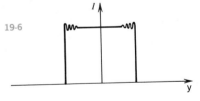

Light of wavelength $\lambda = 6 \times 10^{-7}$ m is incident normally on a single slit 19-7
of width $w = 0.02$ m. (*a*) How far from the plane of the slit should a screen
be placed so that there is a minimum at the center of the screen? (*b*) What
is the width of the central fringe if the screen is placed 1.5 m beyond the
slit? (*c*) How will your numerical answers to these two questions change
if λ is reduced by a factor of 2? (*d*) If w is reduced by a factor of 2?

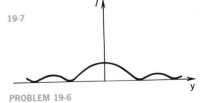

PROBLEM 19-6

19-8   The output of a particular laser on Earth pointed toward the moon is a well-defined beam of light of wavelength $\lambda = 5 \times 10^{-5}$ cm and width (on Earth) $w = 0.1$ cm. (*a*) Make a rough estimate of the angular width of this beam as it approaches the moon. (*b*) Approximately how far away from the output aperture of the laser must one move before this divergence of the beam can be detected? (*c*) How large a spot on the surface of the moon 240,000 miles away would this laser illuminate? (See Fig. 17-1.)

19-9   Describe *qualitatively* what you expect for the Fraunhofer diffraction pattern (*a*) for a single rectangular hole and (*b*) for a single circular hole.

19-10   What is the Fraunhofer interference pattern for a random array of rectangular holes all of the same size, shape, and orientation?

19-11   Repeat Problem 19-7 for a slit of width 0.2 mm.

ELECTROMAGNETIC WAVES

That light consists of waves is supported by the effects examined in the last few chapters, but we can no longer postpone the obvious question: Exactly what does this wave disturbance consist of? What is waving? Why does the wave always move through a vacuum with the constant speed c? In this chapter we shall come to the remarkable conclusion that light is, in fact, an electromagnetic disturbance; it consists of electric and magnetic fields traveling through space. We already have the wherewithal to understand this electromagnetic nature of light: Faraday's law of induced electric fields and Maxwell's law of induced magnetic fields. After deriving the principal properties of electromagnetic waves, including a computation of the speed of light from the fundamental electric and magnetic constants, we shall analyze briefly some of the ways in which light can be created (emitted) or absorbed. Finally, we discuss the momentum carried by a light wave and compute the radiation force, or pressure of light, acting upon a material which reflects or absorbs light.

Light as an electromagnetic phenomenon derived from Faraday's law and Maxwell's law

20-1 THE CONCEPT OF AN ELECTROMAGNETIC WAVE

We have already discussed how an induced electric field **E** is created by a changing magnetic flux. This effect is described in Faraday's law, which was written (see [16-9] and [16-1b]) as

A changing magnetic flux produces an electric field.

$$\Sigma E_{\parallel} \, \delta s = \frac{\Delta \Phi_B}{\Delta t} \qquad [20\text{-}1]$$

where the magnetic flux, defined by [16-7], is $\Phi_B = \Sigma B_{\perp} \, \delta A$. When the magnetic flux through a loop of conducting wire changes with time, the induced electric field **E** drives the charges around the loop. But even in the absence of an actual conducting wire, a changing magnetic field **B** can create an electric field **E** in space.

Maxwell predicted an effect inverse to that discovered by Faraday: an induced magnetic field **B** is created by a changing electric flux Φ_E, described by the law (see [16-11])

A changing electric flux produces a magnetic field.

$$\Sigma B_{\parallel} \, \delta s = \frac{k_m}{k_e} \frac{\Delta \Phi_E}{\Delta t} \qquad [20\text{-}2]$$

where the electric flux Φ_E is defined in Section 16-6 as $\Phi_E = \Sigma E_{\perp} \, \delta A$. The constant k_e is the fundamental electric constant, which in the mks system has the value 9.00×10^9 N-m^2/C^2 in vacuum, and k_m is the fundamental magnetic constant. The unit of current (and hence charge) in the mks system is so defined that k_m has exactly the value 10^{-7} N/A^2 in vacuum (see Section 15-9). The constant

k_e (sometimes written $1/4\pi\epsilon$) gives a measure of the relative strength of the electric interaction in various media, and k_m (sometimes written $\mu/4\pi$) gives a measure of the relative strength of the magnetic interaction in various media.

Taken together, the two induction laws establish a reciprocity between the electric and magnetic fields. It is in working out in some detail the consequences of this relationship that we come to the possible existence of electromagnetic waves.

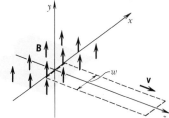

FIGURE 20-1 A magnetic field **B** moving with constant velocity **v** in the positive z direction. Everywhere behind the initial wavefront (an xy plane) the field is uniform; i.e., its magnitude is everywhere B and its direction is everywhere that of the positive y axis. The imaginary rectangular loop of width w used in the application of Faraday's law is located in the xz plane with its width parallel to the x axis and its length parallel to the z axis.

Suppose first that a magnetic field **B** moves through empty space with the constant speed v along the positive z direction as shown in Fig. 20-1. This magnetic field has a constant magnitude B, points in the positive y direction, and extends as an infinite strip behind the xy plane at $z = 0$.

A self-generating magnetic field

How one would produce such a magnetic field will not concern us in this section; we simply assume that such a field can be established and then see that as it moves, it is *self-generating* and must move at a constant noninfinite speed. For simplicity we assume that the leading edge of the magnetic pulse reaches the xy plane at $z = 0$ at the time $t = 0$. An imaginary rectangular loop (shown with dashed lines in Fig. 20-1) lies in the xz plane. Its left end is parallel to the x axis and has the width w; its sides are parallel to the z axis; and its right end lies very far out the positive z axis. At time $t = 0$ the leading edge of the magnetic field is just entering the loop, and after a time Δt has elapsed, this front will have advanced a distance $v\,\Delta t$ along the z direction. Moreover, during this same time interval the magnetic flux through the loop will have increased from its initial zero value by an amount

$$\Delta\Phi_B = \Sigma B_\perp\, \delta A$$

or

$$\Delta\Phi_B = B(wv\,\Delta t)$$

Therefore,

$$\Sigma E_\parallel\, \delta s = \frac{\Delta\Phi_B}{\Delta t} = Bwv \qquad\qquad [20\text{-}3]$$

It is not hard to see that the electromotance, as given by [20-3] comes entirely from the *left* end of the loop along the x axis. The right end is so far away that we may safely assume nothing happens there. Furthermore, there can be no net contribution from the two sides parallel to the z axis because if there were z components to **E**, then for any point at a given value of z on one side there would, *by* *symmetry,* be a corresponding point on the other side where the z component of **E** would be the same. But as we traverse the loop, we go toward positive z on one side of the loop and necessarily toward negative z on the other side. Thus, the contributions from the sides, if any, would cancel. There remains only the left end of the loop, where an electric field **E** must exist *parallel to the x axis.* Let us first establish the magnitude of **E** and then its sense along the x axis. The above reasoning allows us to write

$$\Sigma \mathsf{E}_\| \, \delta s = E w$$

Substituting this result in [20-3] then yields

[20-4] $E = Bv$

This relation informs us that the *magnitude of* **E** *is larger than that of the accompanying* **B** *by the factor v,* the speed with which the magnetic field is traveling. The electric field exists at *all* points within the region in which there is a magnetic field. We can see this by noting that the left end of the imaginary rectangular loop can be shifted around anywhere in the region of the magnetic pulse without changing our argument. Moreover, where B is zero, the induced E is also zero. Thus, an electric field accompanies the magnetic field, both fields traveling together at the speed v.

An electric field accompanies the magnetic field.

We can determine the sense of the induced electric field along the x-axis by applying Lenz' law to the imaginary loop. As the magnetic field enters the loop, the magnetic flux through it increases in the positive y direction (upward). An induced current would be in the direction to oppose this magnetic-flux change; i.e., an induced current would itself generate a magnetic field toward the negative y direction (downward). It follows from the right-hand rule connecting the current direction with the sense of the magnetic field that a current in the end of the rectangle along the x axis, hence the electric field there, would be toward positive x.

Figure 20-2 shows the combined electric and magnetic fields. Both are uniform over the strip. The electric and magnetic fields are at right angles to each other and also at right angles to the direction in which the fields are advancing.

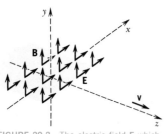

FIGURE 20-2   The electric field **E** which, according to Faraday's law, must be associated with the moving magnetic field of Fig. 20-1.

Equation [20-4] gives a relation between the magnitudes of **E** and of **B**, but it does not specify the speed v. To find this speed we must have still another relation between **E** and **B**. We can arrive at such a relation by applying Maxwell's law of induction [20-2]. Our procedure now will be just the inverse of that used earlier: we start with an *electric* pulse and find that a magnetic pulse must accompany it (see Fig. 20-3). Here we have a constant field **E** along the positive x direction moving at the speed v toward the positive z direction. As before, we consider the effects on an imaginary loop of width b, now one lying in the yz plane. The leading edge of the electric field enters the left end of the loop at time $t = 0$, and after a time Δt has elapsed, we find it to have advanced into the loop a distance $v\,\Delta t$. At the same time the electric flux Φ_E through the loop is increased from zero by the amount

$$\Delta\Phi_E = \Sigma E_\perp\,\delta A = Ebv\,\Delta t$$

Therefore, from [20-2]

$$\Sigma B_\parallel\,\delta s = \frac{k_m}{k_e}\frac{\Delta\Phi_E}{\Delta t} = \frac{k_m}{k_e}Ebv \qquad [20\text{-}5]$$

As before, we find in evaluating the left side of [20-5] that the only contribution comes from the left end of the loop.

This contribution is $\Sigma B_\parallel\,\delta s = Bb$. Then

$$Bb = \frac{k_m}{k_e}Ebv$$

or

$$B = \frac{k_m}{k_e}Ev \qquad [20\text{-}6]$$

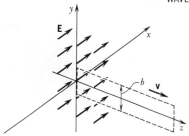

FIGURE 20-3 An electric field **E** moving with constant velocity **v** in the positive z direction. Everywhere behind the initial wavefront (an xy plane) the field is uniform; i.e. its magnitude is everywhere E, and its direction is everywhere that of the positive x axis. The imaginary rectangular loop of width b used in the application of Maxwell's law is located in the yz plane with its width parallel to the y axis and its length parallel to the z axis.

We find the direction of the induced magnetic field by using the right-hand rule for magnetic fields. Recall that when a changing electric flux creates a magnetic field, the effects of the electric flux are like those of a material current whose direction is the same as the direction of the electric-flux *change*. Thus, if our right-hand thumb points in the direction of the electric-flux change, the curved right-hand fingers give the sense of the associated magnetic field. When applied to Fig. 20-3, this implies precisely the relationship of the directions of the **B** and **E** fields which we established earlier (Fig. 20-2). Starting with a magnetic field, we found an associated electric field; starting with an electric field we find an associated magnetic field. One can't exist without the other. Moreover, we have found the relative directions of **E** and **B** and the direction of **v**, the advance of the electromagnetic disturbance, to be consistently the same. We now have two equations, [20-4] and [20-6], which we assume involve the same magnitudes of **E** and **B** and speed of propagation v. Eliminating E and B between these equations, we find

A magnetic field accompanies the electric field.

A self-generating electric field

The speed of an electromagnetic disturbance computed from the electric and magnetic constants

$$v^2 = \frac{k_e}{k_m} = \frac{9.00 \times 10^9 \text{ N-m}^2/\text{C}^2}{10^{-7} \text{ N/A}^2}$$

$$= 9.00 \times 10^{16} \text{ m}^2\text{-A}^2/\text{C}^2$$

But 1 A $= 1$ C/s, so that the units of the above relation check and it can be written

$$v^2 = 9.00 \times 10^{16} \text{ m}^2/\text{s}^2$$

or

$$v = 3.00 \times 10^8 \text{ m/s} = c$$

This is the speed of light! The speed with which an electromagnetic disturbance travels through a vacuum depends only on the fundamental constants k_e and k_m of the electric and magnetic interactions. At last we can say that light is an electromagnetic disturbance traveling through matter-free space at the unique speed 3.0×10^8 m/s, a speed which can be *computed* directly from the laws of electromagnetism and which is, of course, in complete agreement with the measured value—not just to the three significant figures we have quoted but to the full precision permitted by the best independent values of k_e, k_m, and c.

EXAMPLE 20-1

An electromagnetic disturbance, or pulse, is traveling *east*. The *electric* field at one point in space and one instant of time has a magnitude 3.0 N/C and a direction *north*. What are the instantaneous direction and magnitude of the *magnetic* field at this point?

SOLUTION

From Fig. 20-2 we see that if we identify the direction east (that of wave propagation) with that of the positive z axis and north (that of **E**) with the positive x axis, the direction of **B** must be along the positive y axis, or *up*. From [20-4]

$$B = \frac{E}{c} = \frac{3.0 \text{ N/C}}{3.0 \times 10^8 \text{ m/s}} = 10^{-8} \text{ T}$$

Although the electric field is of moderate strength, the associated magnetic field is very feeble indeed.

Query Imagine an electromagnetic disturbance like the one in Example 20-1 but with the electric field directed *south*. If the magnetic field points *up*, in what direction does the disturbance move? If the magnetic field points *down*, in what direction does the disturbance move?

20-2 THE ELECTROMAGNETIC SPECTRUM

We have seen that a pulse of electric field in empty space must be accompanied by a magnetic-field pulse, the two transverse fields moving together at the speed c. Suppose now that the electric field does not change abruptly from zero to some constant value at the leading edge of the pulse and then back to zero again at the trailing edge of the pulse but varies continuously in magnitude from one point in space to another. Such a continuously varying electric field introduces no new ideas, since it is equivalent to a field that changes abruptly in a large number of small steps, as shown for a given point in space in Fig. 20-4. For example, an electric field varying sinusoidally in space is shown along one line in Fig. 20-5. This is a snapshot showing at some one instant how the magnitude of the electric field changes continuously from point to point along the direction of wave propagation. (We know, of course, that at these locations in space the electric field will also vary sinusoidally in time; but it would take a motion picture rather than a snapshot to show this.) If the electric field varies, so must the accompanying magnetic field. Indeed, from [20-4], the magnitude of **B** must always be E/c at any point in space. Furthermore, when the direction of **E** is reversed, so is the direction of the transverse magnetic field. Only then will the electromagnetic wave move along the same direction as before. Thus, the sinusoidally varying **E** field is accompanied by a sinusoidally varying **B** field; both fields attain maximum magnitudes simultaneously, and when **E** is instantaneously zero, so is **B**.

Visible light, i.e., electromagnetic radiation to which the human eye is sensitive, is but one very small portion of the entire electromagnetic spectrum. Figure 20-6 indicates several types of electro-

Sinusoidally varying electromagnetic waves

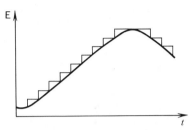

FIGURE 20-4 A continuously varying electric field approximated by a field that changes abruptly in a large number of small steps.

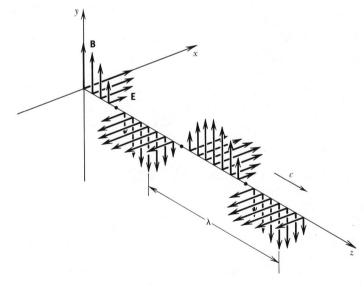

FIGURE 20-5 A sinusoidal electromagnetic field moving with speed c in the positive z direction. Where the magnetic field **B** is in the positive y direction, the electric field **E** is in the positive x direction; and where **B** is in the negative y direction, **E** is in the negative x direction. Everywhere the ratio of the magnitudes is $E/B = c$.

magnetic waves, ranging from the longest-wavelength radio waves to the very short gamma rays. The wavelength and frequency of each kind of radiation in vacuum are connected (see [18-4] and [18-5]) by

$$\lambda = cT = \frac{c}{f}$$

All these waves are alike in the sense that they consist of transverse electric and magnetic fields traveling through a vacuum at the same speed c. They differ immensely in their origin and in their effects on materials. For example, x-rays come from the rearrangement of the innermost electrons in atoms; and most materials are transparent, to a greater or lesser degree, to this relatively short wavelength radiation. On the other hand, infrared (or so-called "heat") radiation originates in the main from molecular vibration, and almost all materials are opaque to this radiation.

+ **20-3 EMISSION AND ABSORPTION OF ELECTROMAGNETIC WAVES**

We have demonstrated that the existence of electromagnetic waves is perfectly consistent with all we know about electric and magnetic fields. How is it possible to create fields satisfying the necessary conditions, a magnetic field **B,** and an electric field **E** of magnitude $E = cB$ perpendicular to **B?** Such a combination of fields will propagate themselves automatically, as described by Faraday's and Maxwell's laws, in a direction mutually perpendicular to **B** and **E.** They detach themselves from their sources, charges and currents, and fly off into empty space on their own.

Electromagnetic waves produced by accelerating electric charges

In nature, the radiation of an electromagnetic disturbance is always associated with the *acceleration* of charges. To describe this process is one of the very difficult theoretical problems of classical physics, and a rigorous treatment is beyond the scope of this book, but a simplified argument can give some physical insight into how an accelerated charge can create the electric and magnetic fields necessary for radiation.

Suppose that an electric charge q is first fixed in position on a space station at rest with respect to some distant fixed star. The static electric field surrounding it falls off inversely as the square of the distance from the source charge, but this static field is zero only at an infinite distance. At the distant fixed star, say 10 light-years[1] out along the positive x axis, a test charge q' will experience a very feeble electric force acting on it (see Fig. 20-7a). Imagine now that the

[1] A light-year is an astronomical unit of distance equal to the distance traveled by light in vacuum in 1 Earth year:

1 light-year
= (186,000 miles/s)(1 year)
$\simeq$ (186,000 miles/s)(1 year)(3.15 $\times$ 10^7 s/year)
$\simeq$ 5.85 $\times$ 10^{12} miles.

source charge is suddenly moved from its first location. In particular, let us assume that this charge starts from rest and moves for a time τ in the positive z direction as shown in Figs. 20-7 and 20-8. It has constant acceleration $+a$ during the first half of the period and a constant deceleration $-a$ of the same magnitude during the second half. At the end of the interval the particle is again at rest but displaced a short distance L along the positive z axis. The position, velocity, and acceleration as functions of time are shown in Fig. 20-8.

We assume that the charge on the space station has been at rest, at least with respect to the frame of reference of the station and the star, for more than 10 years. Just before the time $t = 0$, that is, just before we move the source charge, the feeble Coulomb electric force acting on the charge located on the distant star has the direction shown in Fig. 20-7a. We have drawn in this figure one electric field line, the one pointing directly from the source charge q to the test charge q'. We also know from our previous study of static fields and our recent conclusion about electromagnetic disturbances that at times after $t = \tau + 10$ years the electric force and the electric field line from the source charge to the test charge must have the directions shown in Fig. 20-7c.

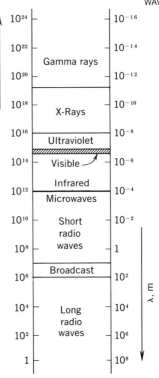

FIGURE 20-6 A portion of the electromagnetic spectrum ranging from long-wavelength (low-frequency) radio waves to short-wavelength (high-frequency) gamma rays.

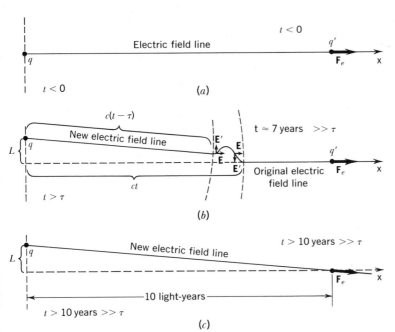

FIGURE 20-7 (a) A particle of charge q at rest on a fixed space station producing an electric force F_e on a particle of charge q' on a distant fixed star. (b) At a time t after q starts to move to' its new position there is still no change in the electric field or the force on an electric charge at distances farther than ct from the original position of q, but at this same time t the electric field has its new direction at all distances less than $c(t - \tau)$, where τ is the time it takes to move q to its new position. (c) Eventually, the electric field has its new direction even as far away' as the distant fixed star. The bump in the electric field line of (b) can be analyzed using Fig. 20-9.

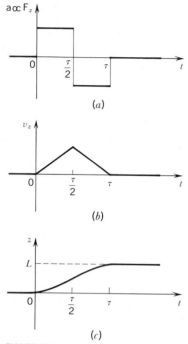

FIGURE 20-8 (a) Acceleration, (b) velocity, and (c) position of the particle of charge q in Fig. 20-7 plotted as a function of time from the time ($t = 0$) q starts to move until the time ($t = \tau$) that it ends its motion and is at its final position a distance L from its origin.

But what happens between times $t = 0$ and $t = 10$ years? If electromagnetic disturbances can travel only with the speed of light, then at some time t between 0 and 10 years the direction of the electric field for distant locations on the x axis, corresponding to $x > ct$, must still be the same as shown in Fig. 20-7a. But at this time the electric field at nearby locations, where $x < c(t - \tau)$, is the same as shown in Fig. 20-7c. It is more difficult to describe the electric field direction for locations near the x axis corresponding to the special region $c(t - \tau) < x < ct$. This part of the field has its origin in the accelerating charge, and previously we have described only the electric field associated with stationary charges. Suffice to say that the detailed analysis of the accelerating charge reveals there is a transverse component of the electric field in this special region, i.e., a component perpendicular to the x axis. Indeed, just drawing a continuous electric field line in this region by connecting the two known portions $x > ct$ and $x < c(t - \tau)$ with a bump, as is done in Fig. 20-7b, already suggests the existence of the transverse component.

In fact, there is a good reason why we should expect to find a transverse electric component in this region: it is precisely the region in which there is a magnetic field! Between time $t = 0$ and $t = \tau/2$ a charge is moving in the $+z$ direction and increasing its speed. At nearby points on the x axis (by nearby we mean $x \ll c\tau/2$), we have a magnetic field increasing into the paper (the $+y$ direction) (see Fig. 20-9a). Using an analysis similar to that used in Section 20-1, we can therefore conclude from Faraday's law and Lenz' law that there is an induced electric field E' in this region in the negative

FIGURE 20-9 Imaginary rectangular loop of width w used in the application of Faraday's law to account for the transverse electric field component in the (a) front and (b) rear of the bump in the electric field line in Fig. 20-7b.

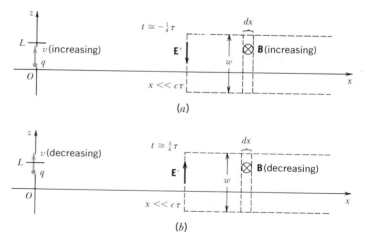

z direction. But a short time later, between $\tau/2$ and τ, the source charge is still moving in the direction of the positive z axis, but now it is slowing down. The magnetic field near the source charge $(x \ll c\tau/2)$ still is pointed into the paper $(+y$ direction), but now it is decreasing in magnitude. As a result, Faraday's law and Lenz' law (see Fig. 20-9b) still give a transverse E' but now in the positive z direction. This change in direction is consistent with the way the slope of the bump changes in Fig. 20-7b.

This transverse (or nonradial) electric field is associated with just the right magnetic field for the combination to propagate along the Coulomb field line drawn in Fig. 20-7a. In so doing, it converts Fig. 20-7a into Fig. 20-7c. It can be shown (see Example 20-2) that the transverse electric field is *proportional to the acceleration* of the charge and *inversely proportional to the distance x* along the x axis. Note that this component of the electric field decreases less rapidly with distance than the radial Coulomb field, which varies inversely with the *square* of the distance x. The greater the acceleration, the greater the transverse radiation field. And the further the radiation propagates, the greater the relative value of the induced transverse field compared to the radial Coulomb field. Indeed, in Example 20-2, we find that at time t

Radiation from an accelerated charge

$$\frac{E'_{trans}}{E_{C,rad}} = 4\frac{t}{\tau}\frac{\langle v \rangle}{c} \qquad\qquad [20\text{-}7]$$

where $\langle v \rangle$ is the average speed of the source charge in the time interval τ. For times comparable to τ, the transverse component is relatively small because our source charge moves with a speed much less than the speed of light. But for times large compared to τ, that is, for t/τ comparable to or larger than $c/\langle v \rangle$, the transverse component will be comparable to or larger than the surviving Coulomb field. It is in this later stage that the radiation field can be said to have broken away from the source charge and its static field.

When a charged particle is set into oscillation by an external force, it accelerates continuously and radiates a sinusoidal electromagnetic wave whose frequency is that of the oscillating charge. If the work done on the charges does not appear in an ever increasing amplitude of oscillation, energy must leave the charge continuously, carried by the electromagnetic wave. Absorption is just the reverse. When an electromagnetic wave impinges upon a charged particle, the incident transverse electric field acts on it, setting it into oscillation, and the work done on the charged particle by the electric field represents energy removed from the incident wave. (We shall discuss the effect of the magnetic field on a charged particle in the next section. We merely remark here that since a magnetic force can do no work, the charged particle loses or gains energy only through the effects of the electric field.)

Any material at a temperature above absolute zero has its component charged particles in thermal motion. This means that electrons undergo accelerations and can radiate energy. If the temperature of the material is high enough, we see some of this radiation when the material becomes *red* hot. At lower temperatures the radiation lies mainly in the *infrared* (longer wavelengths than red) portion of the electromagnetic spectrum. Conversely, when light or other electromagnetic radiation strikes an absorbing material, the electrons gain energy from the oscillating electric fields; the absorbed energy may appear as a rise in the temperature of the absorbing material. (To be sure, the positive charges are also acted upon by the electric field, but these particles, much more massive and more tightly bound than the free electrons, do not undergo appreciable displacements.)

Emission and absorption of electromagnetic radiation

What happens when light is *reflected* from a material? The incident beam sets electrons in oscillation, and these oscillating charges act as new radiators. Thus, an electromagnetic wave undergoing reflection at a surface is effectively absorbed by atoms at the surface, which then reradiate the energy as a reflected wave.

c EXAMPLE 20-2   Show that the induced electric field E' in Fig. 20-9a is given by

$$E'_z = -k_m \frac{qa}{x}$$

where a is the acceleration of the source charge q. Use this result to derive [20-7].

SOLUTION   Consider the narrow rectangular path extending along the x axis in Fig. 20-9a. The magnetic field due to the moving charge is into the paper ($+y$ direction) with the magnitude (see [15-10])

$$B \approx k_m \frac{qv \sin 90°}{x^2}$$

where v is the speed of the source and x is the distance from the source to a point within the narrow loop. Because all points within the narrow loop of width w are near the x axis, and because the source charge always remains near the origin on the x axis, the angle θ between the velocity of the source particle and the direction to any point within the rectangular loop is very nearly $90°$. Thus the magnetic flux through the loop is

$$\Phi_B = \int_x^\infty k_m \frac{qv}{x^2} w \, dx = k_m \frac{qv \, w}{x}$$

The induced electromotance around the loop is

$$\Sigma \mathsf{E}'_\parallel \delta s = \frac{d\Phi_B}{dt}$$

or

$$\dot{E}'_z w + 0 = \frac{k_m qw}{x} \frac{dv}{dt}$$

Here we have assumed that the end of the loop at $x = \infty$ is so far away that there can be no induced field there and that any contribution to the electromotance from one long side parallel to the x axis must, by *symmetry*, be cancelled by an equal but opposite contribution as we move in the opposite direction along the opposite side. Therefore we have

$$E'_z = k_m \frac{qa}{x}$$

where $a = dv/dt$ is the acceleration of the source charge. We can use Lenz' law to obtain the sign of E'_z. For times between 0 and $\tau/2$ the source charge is speeding up, the current is increasing, and the magnetic field is increasing into the paper. Thus the sign of E'_z is negative; i.e., it would cause current to move around the loop in a sense so as to create a new flux opposing the original *change* in flux. Between $\tau/2$ and τ, the direction is reversed.

During the time the radiation field is passing a given location x,

$$E'_z = E' = k_m \frac{qa}{x} \qquad \text{(radiation field)}$$

while the static Coulomb field is always very nearly

$$E_x = E_c = k_e \frac{q}{x^2} \qquad \text{(Coulomb field)}$$

Therefore

$$\frac{E'}{E_c} = \frac{k_m}{k_e} ax = \frac{ax}{c^2}$$

But $\langle v \rangle = \frac{1}{2}(v_i + v_f) = \frac{1}{2}(0 + a\tau/2) = a\tau/4$ is the average speed of the source charge during both the initial $\tau/2$ s when the particle accelerates from rest and the final $\tau/2$ s when the particle decelerates to rest. Finally, with $x = ct$

$$\frac{E'}{E_c} = \frac{4\langle v \rangle}{\tau} \frac{x}{c^2} = \frac{4\langle v \rangle}{c} \frac{t}{\tau}$$

+   **20-4 RADIATION FORCE
    AND THE MOMENTUM
    OF LIGHT**

Some delicate experiments show that light striking a surface produces a force upon it (see Section 17-3). It is now easy to show that this radiation force has its origin in the *magnetic* field of an electromagnetic wave.

Assume that a sinusoidally varying electromagnetic beam impinges on a material along a direction normal to the surface. We suppose the material to be black; it absorbs the wave completely. An electron in the material is driven in oscillation by the wave's transverse electric field. The energy carried by the incoming electromagnetic wave appears as work done on the oscillating charge by this electric force, and the absorbed energy is then distributed as thermal energy to nearby atoms. It can be shown that as the electron oscillates, the electric force on it at each instant is (very nearly) along the same direction as the electron's velocity; i.e., the electron always moves in the direction of the electric force driving it in oscillation. As a result, energy is always absorbed from the wave.

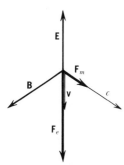

FIGURE 20-10 The relative directions of the electric field **E**, the magnetic field **B**, the wave velocity **c**, the electric force F_e on the free electrons at the surface, and the resulting electron velocity **v** and magnetic force F_m as an electromagnetic wave is incident normally on the surface of a metal.

Figure 20-10 shows the electric and magnetic fields of the incident beam at one instant acting on a moving electron. The electric force F_e has magnitude eE and is opposite in direction to **E**. The electron's velocity **v,** which is parallel to **E**, is at right angles to **B**; therefore the magnetic force F_m has magnitude (see [15-12])

[20-8] $F_m = evB$

Moreover, as Fig. 20-10 shows, the direction of F_m is *along* the direction of wave propagation, i.e., the direction of the wave velocity. This is true even when **v** changes direction, since both **E** and **B** change direction along with it. The *magnetic* force is the radiation force pushing on the absorbing material!

An electromagnetic wave produces a radiation force on an absorber.

Now let us compute the magnitude of this magnetic force. We can use [20-4] to write [20-8] as

[20-9] $F_m = evB = ev\dfrac{E}{c} = eE\dfrac{v}{c} = F_e\dfrac{v}{c}$

461

+ SECTION 20-4
RADIATION
FORCE AND
THE MOMENTUM
OF LIGHT

The force $\mathbf{F}_e$ does work *on the electron* because the electron moves at a velocity $\mathbf{v}$ in the same direction as $\mathbf{F}_e$. Since $\langle v \rangle = \Delta x / \Delta t$,

$$F_e \langle v \rangle = \frac{F_e \Delta x}{\Delta t} = \frac{\Delta W}{\Delta t} = \langle P \rangle \qquad [20\text{-}10]$$

where ΔW is the work done by the electric force in the time Δt and by definition, the rate of doing work is the power P. Thus, [20-9] can be written

Complete absorption: $F_m = \dfrac{F_e v}{c} = \dfrac{P}{c}$ $\qquad$ [20-11]

The power P is the rate at which the electric force does work on the electron in a completely absorbing material. It is then also the rate at which energy is extracted from the incoming electromagnetic wave. Therefore, in the case of complete absorption, we can identify P with the power of the beam, i.e., the energy of the beam crossing any fixed transverse area of the beam per unit time.

In short, when an electromagnetic wave strikes a completely absorbing material, the electric field does work on charged particles and sets them in motion, while the magnetic field, acting on the moving charges, produces a radiation force on these charges (and therefore also on the material in which they are embedded) along the direction of the wave propagation.

EXAMPLE 20-3

What is the radiation force of a 30-W beam of light striking a black surface along the normal?

SOLUTION

From [20-11]

$$F_m = \frac{P}{c} = \frac{30 \text{ W}}{3.0 \times 10^8 \text{ m/s}} = 10^{-7} \text{ N}$$

The force is very small indeed. It would take a beam of 300,000 kW to produce a force of only 1 N. Note that the frequency of the wave does not enter in [20-11]. A completely absorbed 30-W beam of radio waves or of x-rays would produce exactly the same radiation force.

EXAMPLE 20-4

With a bright sun overhead, the total electromagnetic radiation from the sun striking an area of 1 m^2 at the Earth's surface is 1,400 W. What is the radiation pressure of sunlight?

SOLUTION

We find the radiation force on the 1-m^2 area from Eq. [20-11]

$$F_m = \frac{P}{c} = \frac{1,400 \text{ W}}{3.0 \times 10^8 \text{ m/s}} = 4.7 \times 10^{-6} \text{ N}$$

The radiation pressure is just the radiation force F_m per unit transverse area A

$$\frac{F_m}{A} = \frac{4.7 \times 10^{-6} \text{ N}}{1.0 \text{ m}^2} = 4.7 \times 10^{-6} \text{ N/m}^2$$

This very small radiation pressure is to be contrasted with the pressure of the Earth's atmosphere, approximately 10^5 N/m^2.

(a) Absorption

(b) Emission

FIGURE 20-11 The direction of the radiation force F_r when an object (a) absorbs or (b) emits electromagnetic radiation.

A radiation force also acts on a material emitting electromagnetic waves. For emission, the electrons in the material again undergo oscillations, but now the energy originally residing in the emitting material is carried away as the energy of an outgoing beam. It is just as if the absorption process were imagined to run backward in time. Again there is a radiation force on the material, but now it is a *recoil* force acting in the direction opposite to that of the emitted wave. The power P denotes the power emitted (see Fig. 20-11).

The radiation force for complete reflection

Reflection of light by a mirror is merely absorption of an incident wave accompanied by the re-emission of the same wave. Since a radiation force of magnitude $F_m = P/c$ acts in both absorption and in emission, we have for the force on the mirror when incidence is normal

[20-12] Complete reflection: $F_m = \dfrac{2P}{c}$

The direction of the radiation force is the same as that of the incident wave.

EXAMPLE 20-5

A powerful flashlight of 1 kgm mass emits a beam of 300 W. The beam is turned on for 10 s. (a) What is the recoil force acting on the flashlight while it is turned on? (b) What is the force of this beam on a mirror which it illuminates? (c) If the flashlight is initially at rest but free to coast, what is the velocity acquired by the flashlight after the beam has been turned on and then off?

SOLUTION

a The recoil force on the flashlight is

$$F_m = \frac{P}{c} = \frac{300 \text{ W}}{3.0 \times 10^8 \text{ m/s}} = 10^{-6} \text{ N}$$

463

+ SECTION 20-4
RADIATION
FORCE AND
THE MOMENTUM
OF LIGHT

b The radiation force on the completely reflecting surface of the mirror is

$$F_m = \frac{2P}{c} = 2 \times 10^{-6}\,\text{N}$$

c The flashlight is actually a very simple form of rocket. When light is emitted in one direction, a radiation force acts in the opposite direction to set the light emitter in motion and change its momentum. In general,

$$F = \frac{\Delta p}{\Delta t}$$

where Δp is change in the momentum in the time Δt. Therefore, since the flashlight starts from rest, and $F = P/c$

$$\frac{P}{c} = \frac{mv - 0}{\Delta t}$$

or

$$v = \frac{P\,\Delta t}{c\,m} = \frac{(300\ \text{W})(10\ \text{s})}{(3.0 \times 10^8\ \text{m/s})(1.0\ \text{kgm})} = 10^{-5}\ \text{m/s}$$

The flashlight acquires a speed of a mere $\frac{1}{100}$ mm/s. Of course, if the flashlight beam remained on for a long time, the very small force, acting continuously, would be able to accelerate an isolated flashlight to a high speed. Note again that the frequency of the radiation does not enter; an electromagnetic rocket can operate by emitting any type of waves (see Example 22-8).

If light can exert a force, it must also possess momentum, as the following thought experiment will show. Imagine that a light source emits a beam in the east direction which travels to a star 10 light-years away. The emitter recoils toward the west, and 10 years later the beam strikes an absorber, whose momentum is changed toward the east by exactly the same amount as that of the emitter. Initial and final net momenta are therefore equal, but unless we attribute a momentum to the beam itself, we have a violation of momentum conservation during the 10 years in between. The radiation force on the emitter is $F = P/c$, so that the change in momentum of the emitter is

An electromagnetic beam has momentum.

$$\Delta p = \langle F \rangle\,\Delta t = \frac{\Delta W}{c} \qquad\qquad [20\text{-}13]$$

where $\Delta W = \langle P \rangle \Delta t$ is the energy E of the emitted wave. It is now clear from the momentum-conservation law that the momentum carried off by the emitted beam must also have the magnitude Δp. We then write for the momentum p of an electromagnetic beam of energy E traveling along a single direction in space

[20-14] $\quad p = \dfrac{E}{c}$

The momentum of light is merely the energy of the beam divided by the speed c.

EXAMPLE 20-6 A radar antenna emits a pulse of 1.0 megawatt (MW) over a time of 1.0 microsecond (μs). What is the momentum of this pulse?

SOLUTION The energy emitted is $E = (1.0 \times 10^6 \text{ W})(1.0 \times 10^{-6} \text{ s}) = 1.0 \text{ J}$. Therefore, the momentum is from [20-14]

$$p = \frac{E}{c} = \frac{1.0 \text{ J}}{3.0 \times 10^8 \text{ m/s}} = 3.3 \times 10^{-9} \text{ kgm-m/s}$$

An electromagnetic beam has energy, and it has momentum. From our familiarity with Newtonian mechanics we might be inclined to write the energy E as $\frac{1}{2}mv^2 = \frac{1}{2}mc^2$, where the mass m, in some sense not yet clear, is the mass of the electromagnetic wave and $v = c$ is its speed. Similarly, we would be inclined to write the wave's momentum as $p = mv = mc$. Now if we substitute these relations in [20-14] we find a surprising paradox

$$p = \frac{E}{c}$$

$$mc \overset{?}{=} \frac{\frac{1}{2}mc^2}{c}$$

$$1 \overset{?}{=} \tfrac{1}{2}$$

There is something basically wrong here. It cannot be the relation $p = E/c$, which we established quite carefully. But we have never confirmed that the momentum and energy of a particle are given by mv and $\frac{1}{2}mv^2$ for speeds other than the very low speeds of our normal laboratory experiments. Indeed, we shall soon see that these Newtonian relations for momentum and kinetic energy are not valid for objects traveling at speeds close to the speed of light, and certainly not for light itself. The proper relations are given by the special theory of relativity, the topic we begin next in Chapter 21.

The Newtonian relations for momentum and energy do not apply to an electromagnetic wave.

Starting with Faraday's law of induced electric fields and Maxwell's law of induced magnetic fields, one can predict the existence of electromagnetic waves having the following properties:

1 Their speed in vacuum is the same for all energies and all frequencies and has the value $c = \sqrt{k_e/k_m} = 3.0 \times 10^8$ m/s.

2 They consist of a self-generating combination of an electric field and a magnetic field perpendicular to each other and to the direction of propagation of the wave (see Figs. 20-2 and 20-5); the ratio of the magnitudes of these two fields has at each point the value $E/B = c$.

3 These waves originate from *accelerating* charges.

4 When a uniform beam is incident normally upon, and completely absorbed by, a surface, the radiation force on the surface is

$$F_m = \frac{P}{c} \qquad [20\text{-}11]$$

where P is the power of the beam, i.e., the energy crossing a transverse area of the beam per unit time. For complete reflection the force is doubled.

$$F_m = \frac{2P}{c} \qquad [20\text{-}12]$$

5 If E is the energy of the beam, its momentum is

$$p = \frac{E}{c} \qquad [20\text{-}14]$$

20-1 A plane electromagnetic wave is traveling in air. At one instant of time the electric field is vertically upward and the magnetic field is eastward. In what direction is the wave moving?

20-2 The plane electromagnetic disturbance shown in the figure has, at time $t = 0$, an **E** of uniform magnitude pointing in the positive y direction at all values of $z < 0$. This disturbance is moving with speed v in the positive z direction. (a) Use Maxwell's law to demonstrate the existence of a magnetic field **B** associated with the disturbance and to determine the direction of **B.**

20-3 A plane electromagnetic wave is traveling in the positive y direction. At one instant of time the electric field is in the negative x direction with a magnitude 60 V/m. What are the (a) direction and (b) magnitude of the magnetic field at this instant of time?

20-4 Calculate the wavelengths of the electromagnetic disturbances emitted by the following sources which have the frequencies listed: commercial power generator, 60 Hz; WQXR-AM, 1,560 kHz; WQXR-FM, 96.3 MHz; TV Channel 2, 54 to 60 MHz; TV Channel 13, 210 to 216 MHz; microwave radar, 30,000 MHz; gas laser, 4.738×10^{14} Hz.

20-5 An electromagnetic wave carries equal amounts of electric and magnetic energy per unit volume. Derive an expression for the magnetic energy per unit volume in terms of B and k_m. *Hint:* Start with [13-8].

20-6 Show that the energy per unit length along the direction of wave propagation in an electromagnetic beam of power P is P/c.

20-7 A 100-W electromagnetic laser beam is directed straight up from the surface of the Earth. (*a*) How heavy a sheet of perfectly reflecting metal can be floated in this beam at the Earth's surface? (*b*) How heavy a completely absorbing sheet can be floated at the same position? (A 100-W beam is so intense that if it diverges with an angle equal to 1 second of arc, it is still capable of igniting wood over 1 mile away and being detected by a receiver with a 3-in.-diameter lens over 10 billion miles away!)

20-8 What is the radiation force of a 100-W beam of light incident normally on a surface if half of the incident radiation is absorbed and half is reflected?

20-9 A perfectly reflecting radiation "sail" is oriented in interstellar space so that rays from the distant sun are incident at right angles to its mirrorlike surface. At some one position the area of the sail is such that the gravitational attraction of the sail by the sun is exactly balanced by the radiation force from the sun, and the sail floats. Show then that the sail floats at *any* separation distance from the sun (large compared to the sun's diameter) so long as the rays from the sun are incident normal to the surface.

20-10 A powerful laser emits a pulse of light whose energy is 5 J. (*a*) What is the momentum of the pulse? (*b*) If the laser is on for only 10^{-6} s while emitting the pulse, what is the force of the pulse on a material that is absorbing it completely? (*c*) If the laser emits a pulse as given above every 10^{-2} s, what is the average force on a material absorbing these pulses?

ANSWERS
TO SELECTED
PROBLEMS

CHAPTER 1 1-1 (a) 10.0 m/s, (b) 32.8 ft/s, (c) 22.4 mph 1-3 (a) 2.16×10^{11} m,
(b) 1.34×10^8 miles 1-5 (a) 60 mph, (b) 60 mph 1-9 25 mph
1-11 (a) 0, (b) 100 m, (c) 0, (d) 1.89 m/s 1-13 (a) 25 mph,
(b) 12.5 mph 1-15 (a) $[2\pi A_0/T_0] \cos (2\pi t/T_0)$, (b) 0, (c) $2\pi A_0/T_0$
1-17 2.0 m/s 1-19 3.4 m/s

CHAPTER 2 2-1 2 m/s 2-3 (a) 6.0 kgm-m/s, (b) 0.2 m/s 2-5 0.5 m/s to right
2-7 (a) 9.1 m/s, (b) no; the two bullets are not an isolated system
2-9 1.2×10^5 cm/s 2-11 (a) 1.6 m/s to left, (b) 4.0 m/s to left
2-13 (a) 7.5 m/s, (b) 16.5 m/s, (c) 13.5 m/s 2-15 (a) 0.6 m
displacement, (b) 0.4 m displacement, i.e., no change 2-17 27 km/s

CHAPTER 3 3-1 20 N 3-3 (a) 107 N, (b) the ground 3-5 −12 N
3-7 (a) 160 cm/s, west, (b) 360 cm/s, west 3-9 (a) 5 N to left, (b) 5 N
to left, (c) 7.5 N to left 3-11 1.6 N upward 3-13 (a) Momentum to
left at A and B, and force to right at B, (b) force to right but no
momentum at C, (c) momentum to right at B and A, and force to right
at B 3-15 (a) 0, (b) 0.98 N 3-17 (a) +, 0, −, 0, +, (b) 0, −, 0, +, 0
3-19 (a) 33 lb, (b) 0

CHAPTER 4 4-1 19.6 m/s 4-3 (a) 17 m/s, (b) 850 m 4-5 (a) 15, 10, 15, 20,
20 m/s, (b) 25, 30, 25, 20, 20 m/s, (c) 20, 20, 20, 20, 20 m/s
4-7 80 ft/s 4-9 12 miles 4-11 (a) 5.4 miles, (b) 4.7 min
4-13 (a) 323 N, (b) 294 N, (c) 309 N 4-15 (c) 28 N, (d) 70 N
4-17 (b) 2,170 N 4-19 (a) 66 lb, (b) 1.9 lb, (c) 10.1 lb, (d) 12 ft

CHAPTER 5 5-1 (a) 192 miles, 28° east of north, (b) 171 miles, 20.5° east of north
5-3 $0.8v_0$ and $0.3v_0$ 5-5 (a) $y = 16$ m at 10 s, (b) $y = 24$ m at 10 s
5-7 0.97 m/s 5-9 0.69 kgm 5-11 (a) 64 m, (b) 32 m 5-13 50 lb
5-15 (a) attractive, 7.3×10^{-8} N, (b) 8.9×10^{-30} N 5-17 (a) 0,
(b) 2,270 lb central

6-1 0   **6-3** (a) 7.2 m, (b) 7.2 m, (c) 0   **6-5** (a) 4 m/s, (b) 0, (c) 6 kgm is 6 m to east, and 3 kgm is 12 m to west   **6-7** (a) $\frac{4}{3}$ m/s to west, (b) 4 m/s to south, (c) 4 m/s to east, (d) 4 m/s to east   **6-9** (a) 1 m/s to east, (b) 2 m/s to east and 1 m/s to west before, and both at rest after   **6-11** Before collision they approach from opposite directions each with speed $v_0/2$; after collision they separate in opposite directions each with speed $v_0/2$ **6-13** Before collision they approach from opposite directions (east and west) each with speed $v_0/2$; after collision they separate in opposite directions (north and south) each with speed $v_0/2$ **6-15** (a) 30° west of north, (b) 260 mph   **6-17** (a) 0.77 s, (b) 0.77 s, (c) 0.63 s, (d) as long as it takes the elevator to hit the bottom of the shaft

7-1 3.6 kgm-m^2/s   **7-3** (a) 100 kgm-m^2/s, (b) 112 kgm-m^2/s, (c) no, (d) no   **7-5** (a) 2 m from west end; remains at rest, (b) 60 kgm moves south at 1.0 m/s, and 30 kgm moves north at 2.0 m/s, (c) 30 N, (d) 30 N   **7-7** (a) 18 kgm-m/s, (b) no, (c) 3.14 s, (d) man's motion unchanged   **7-9** (a) 48 kgm-m^2/s counterclockwise, (b) 36 N-m clockwise   **7-11** About 33   **7-13** (a) 5.0 N-m, (b) 5.0 kgm-m^2/s, (c) 19 rpm   **7-15** Three-quarters of the way to the back end   **7-17** Yes, 90 lb   **7-19** (a) 50 N-m counterclockwise, (b) 20 N-m clockwise, (c) yes, (d) right

8-1 (a) 0.5 mile, (b) 30 mph   **8-3** (a) 30 m/s, (b) 30 m/s, (c) 20 m/s to west, (d) 10 m/s to east   **8-7** 0.85   **8-9** (a) One ball moves out the opposite end, (b) two balls move together out the opposite end **8-11** $M/(M + m)$ **8-13** (a) Less than 90°, (b) greater than 90° **8-15** (a) +640 N-m, (b) −320 N-m, (c) +320 N-m, (d) 8.0 m/s **8-17** −15 N-m

9-1 (a) 0.072 J, (b) 80 N/m   **9-3** 13 m/s   **9-5** 209 J   **9-7** 53 m/s **9-9** (a) 1.7 m/s, (b) 1.0 m/s, (c) 1.4 m/s   **9-11** 0.69 kgm **9-13** (a) 160 ft/s, (b) 226 ft/s, (c) 179 ft/s   **9-17** 1.8 ft   **9-19** (a) No; the rope under tension does work on the monkey, (b) 25.2 lb

CHAPTER 10   10-1 546°C   10-3 230 m/s   10-5 (a) 1, (b) 1.55   10-7 (a) 3.0 atm,
(b) 3.18×10^4 m/s   10-9 2.5×10^{-14} kgm

CHAPTER 11   11-1 (a) The box starts to move to the right when the marbles collide
with the right side of the box; later when the left side of the box and
the marbles collide, the box comes to rest and the marbles resume their
original motion, (b) Nmv_0, (c) $\frac{1}{2}Nmv_0^2$, (d) although the marbles are in
motion, the box remains at rest except for small random displacements
about its average position, (e) 0, (f) $\frac{1}{2}Nmv_0^2$ 11-3 (a) +, (b) 0, (c) −,
(d) −   11-5 (a) No, (b) no   11-7 0.05°C, (b) 2×10^5 W   11-9 $\frac{1}{10}$ raised to
the power 3×10^{20}

CHAPTER 12   12-1 $\frac{1}{3}$ 12-3 1.53 AU   12-5 3.84×10^7 m, or 23,900 miles
12-7 1.3×10^2 m/s   12-9 $\frac{1}{2}$ (a) 0, (b) F_0 12-13 (a) 0,
(b) 0.098 N, (c) 4.9 N   12-15 (a) 0, (b) $GmM/2R$, (c) GmM/R
12-17 $\frac{1}{2}\sqrt{Gm/r}$

CHAPTER 13   13-1 5.0×10^{-10} C   13-3 4.9×10^5 N   13-5 5.8×10^{-7} C
13-7 (a) 0.75 cm, (b) 1.12×10^{-7} C   13-9 (a) 4.3 N straight up,
(b) 1.08×10^7 N/C straight down   13-11 (a) 1.1×10^{-2} C,
(b) 8.8×10^{-4} C/m^2 13-13 (a) 2×10^7 N/C toward center, (b) 2×10^7
N/C toward center, (c) 0   13-15 (a) unchanged, (b) inversely proportional
to distance from plane surface

CHAPTER 14   14-1 (a) 0.182 J, (b) 2.0×10^{-6} J   14-3 (a) -2.0×10^{-6} J,
(b) -2.0×10^{-6} J   14-5 (a) 3.4×10^2 N/C, perpendicular to the plates
directed from the positive to negative charge, (b) 1.7 V
14-7 Yes, 1.9 eV   14-9 (a) 2.2×10^4 m/s, (b) 1.20×10^{-18} J, or
7.5 eV   14-13 (a) 3.7×10^4 m/s, (b) 0, and 5.2×10^4 m/s
14-15 (a) $q_0(k_e/m_0R)^{1/2}$, (b) $2q_0(k_e/3m_0R)^{1/2}$ and $q_0(k_e/3m_0R)^{1/2}$
14-17 (a) k_eQq/d^2, (b) 0   14-19 (a) 0; k_eq_1/r^2 toward center; $k_e(q_1 + Q_2)/r^2$
toward center, (b) force on free electrons in wire is away from center

15-1 1 cm/s 15-3 (a) 2 Ω, (b) 18 W 15-5 80 A into the paper
15-7 (a) zero, (b) zero, (c) 7.2×10^{-12} N vertically downward,
(d) 7.2×10^{-12} N vertically upward, (e) 4.3×10^{-12} N vertically
downward, (f) 7.2×10^{-12} N eastward, (g) 7.2×10^{-12} N westward
15-9 (a) west, (b) 7.5×10^{-2} N 15-11 (a) 2.9×10^7 m/s,
(b) 8.7×10^{-8} s, (c) 8.6×10^6 V 15-13 (a) west, (b) 4.3×10^{10} m/s
(an impossible speed, since it exceeds that of light; more about this in
Chapter 22), (c) $F_m = 2.9 \times 10^7 F_g$ 15-15 0.2 N/m 15-17 (a) 1 mA,
(b) 13 T 15-19 (b) Perpendicular to the plane of the circular loop

16-1 (a) 10 A, (b) 0.05 Ω, (c) 1.0 V, (d) 15 W 16-3 (a) 3 A, (b) 4 V,
(c) 2 V, (d) 0 V, (e) 0 Ω 16-5 (a) 16 V, (b) 22 V, (c) 54 V
16-7 (a) 0.016 Wb, (b) zero, (c) zero 16-9 (a) 0.06 V, counterclockwise,
(b) 0.19 V, counterclockwise, (c) 0.15 V, counterclockwise
16-11 (a) 0.06 V, (b) a to b, (c) 0.72 mW 16-13 (a) positive z direction,
(b) 0.04 T/s 16-17 (a) 1 Wb/s, (b) 1 V, (c) 0.1 A, (d) counterclockwise,
(e) 0.10 W, (f) 0.05 N, (g) force of hand to left

17-1 counterclockwise 17-3 1.3 s 17-5 (a) 38°, (b) 33°
17-7 $t \sin (\theta_1 - \theta_2)/\cos \theta_2$ 17-9 2.4 ft 17-11 (a) 56°, (b) glass
17-13 760 cm^2

18-1 (a) halved, (b) doubled, (c) doubled 18-3 16.5 mm to 16.5 m
18-7 (a) 0.15 mm, (b) double source frequency, halve D, double d,
double refractive index (e.g., fill space between slits and screen with
material such as zincite, for which $n = 2.013$) 18-9 (a) 100 cm, (b) one
transmitter is oscillating with the opposite polarity relative to the
other 18-11 (a) 10 m, (b) north-south 18-13 (a) Distance of a
particular principal maximum from center is halved, (b) principal
maxima unchanged in position but narrower, (c) separation of principal
maxima is doubled

19-1 10^3 km 19-3 5 m 19-5 3.3×10^{14} Hz 19-7 (a) 83 m,
(b) 0.02 m, (c) distance in (a) doubled and in (b) unchanged,
(d) distance in (a) reduced by factor 4 and in (b) reduced by factor 2
     (a) 0.8 cm, (b) 0.9 cm, (c) doubled, reduced by a factor $\frac{1}{2}$,
(d) reduced by a factor $\frac{1}{4}$, doubled

CHAPTER 20 20-1 North 20-3 (a) positive z direction, (b) 2.0×10^{-7} T
20-5 $B^2/8\pi k_m$ 20-7 (a) 6.7×10^{-7} N, (b) 3.3×10^{-7} N

CHAPTER 21 21-1 6.3×10^{-5} s 21-3 2.3 h 21-5 (a) $0.8c$, (b) 2.5×10^{-8} s
21-7 (a) 35.8 years, (b) 3.4 light-years 21-9 34.7 m 21-11 (a) 117 m,
(b) 42 m, (c) no 21-13 (a) 8 and 24 ft, (b) 8 and 14 ft
21-15 $0.98c$ 21-17 $0.995c$

CHAPTER 22 22-1 (a) 9.1×10^{-31} kgm, (b) 9.1×10^{-31} kgm, (c) 45.5×10^{-31} kgm
22-3 $\frac{1}{2}L_0$ 22-5 $0.87c$ 22-7 (a) 1.56 GeV, (b) 0.62 GeV
22-9 (a) 1 GeV/c, (b) 1.7 GeV/c, (c) 1 GeV/c 22-11 (a) 2.7×10^{-30} kgm,
1.0 MeV, (b) 2.3×10^{-27} kgm, 1 GeV, (c) 5.3×10^{-26} kgm, 29 GeV
22-13 (a) 5×10^7 kgm, (b) no 22-17 (a) 1.0 MeV, (b) 1.4 MeV/c,
(c) 4.7×10^{-3} T

CHAPTER 23 23-1 1.48 eV 23-3 9.9 eV 23-5 7.2×10^{-34} J-s
23-7 (a) 6.2×10^{-9} eV, (b) 4.1×10^{-7} eV, (c) 1.24×10^{-5} eV,
(d) 2.48 eV 23-9 (a) 16×10^8, (b) 6.0×10^{24}, (c) 3.0 MW
(a) 0.050 Å, (b) 0.052 Å, (c) 0.002 Å

CHAPTER 24 24-1 (a) 7.7 MeV, (b) 3.0×10^{-14} m, (c) 7.7 MeV, (d) 1.5×10^{-14} m
24-3 (a) 1.2×10^3 Å, (b) 9.1×10^2 Å at the ionization limit
24-7 Three 24-9 (a) 54.3 eV, (b) 0.264 Å

CHAPTER 25 25-1 6.6×10^{-34} m 25-3 (a) 3.8×10^{-2} eV, (b) 1.5 Å
25-5 (a) 0.12Å, (b) 1.7° 25-9 (a) $\gtrsim 10^{-28}$ m/s $= 10^{-18}$ Å/s,
(b) $\gtrsim 10^7$ m/s 25-11 16

CHAPTER 26 26-1 $\frac{1}{6}, \frac{1}{3}, \frac{1}{2}, \frac{2}{3}, \frac{5}{6}$ Hz 26-5 (n,l,m_l,m_s): $(0,0,0,\pm\frac{1}{2})$, $(1,0,0,\pm\frac{1}{2})$,
$(2,1,0,\pm\frac{1}{2})$, $(2,1,1,\pm\frac{1}{2})$, $(2,1,-1,\pm\frac{1}{2})$ 26-9 (a) $y =$
$2A \cos(n\pi vt/L) \sin(n\pi x/L)$, where $n = 1,2,3$, and 4, (b) $nv/2L$, $2L/n$

27-7 $^{17}_8\text{O}$ 27-9 $\frac{1}{16}$ 27-11 $1/e$ CHAPTER 27

28-1 (*a*) 1.53 MeV, (*b*) 0.26 MeV 28-3 (*b*) 3.06 MeV, (*c*) 0.51 MeV CHAPTER 28
28-5 (*a*) No, momentum, (*b*) yes, (*c*) no, baryon, (*d*) yes, (*e*) yes,
(*f*) no, charge, (*g*) no, electron number, (*h*) no, energy, (*i*) no, electron
number, (*j*) yes, (*k*) yes, (*l*) no, electric charge 28-7 $\sim 10^{-9}$ eV

INDEX